SpringerBriefs in Pharmacology and Toxicology

For further volumes:
http://www.springer.com/series/10423

Vartika Jain · Surendra K. Verma

Pharmacology of *Bombax ceiba* Linn.

 Springer

Dr. Vartika Jain
Department of Botany
University College of Science
Mohanlal Sukhadia University
Udaipur 313003, Rajasthan
India

Dr. Surendra K. Verma
Department of Medicine
Indigenous Drug Research Centre
RNT Medical College
Udaipur 313001, Rajasthan
India

ISSN 2193-4762   e-ISSN 2193-4770
ISBN 978-3-642-27903-4   e-ISBN 978-3-642-27904-1
DOI 10.1007/978-3-642-27904-1
Springer Heidelberg New York Dordrecht London

Library of Congress Control Number: 2011946211

Printed on acid-free paper

Springer is part of Springer Science+Business Media (www.springer.com)

Foreword

The bio-cultural diversity of nature has gifted wide range of resources of therapeutically important ethno-medicinal plants. Presently numerous plants and their molecules have been screened and identified as natural bioactive products, which not only enrich the therapeutic compendium but also provide rich source of new pharmaceuticals, cosmetics, agrochemicals and others. The pharmaceutical industries are working on proving the benefits of various botanicals for our health and nutraceuticals for another rapidly expanding market. I found that this comprehensive treatise on *Bombax ceiba* Linn. has highlighted on the reported activities and traditional claims on this plant, which possesses immense potential as hypotensive, hypoglycemic, hypolipidemic, hepatoprotective, chemopreventive, antioxidant, fibrinolytic etc. The information presented in this volume will have a considerable impact on the pharmacological and therapeutic profile of *Bombax ceiba* Linn. The examples and evidences presented in this book strongly support the ethno-medicinal importance, quality parameters as well as the different screening profiles of this species.

It is thus of great interest to have the present book dealing with the approach that can be followed in the evaluation of therapeutic efficacy of *Bombax ceiba* Linn., which is a timely attempt of Dr. S. K. Verma and Dr. Vartika Jain to write a book on the "Pharmacology of *Bombax ceiba* Linn." I greatly appreciate their efforts and I am sure this will be very useful for all who works on development of evidences based on the claim of traditional medicine as well as to all research workers in the field of natural products.

Pulok K. Mukherjee, M.Pharm, Ph.D, FRSC
Director School of Natural Product Studies
Jadavpur University
Kolkata, India

Preface

Nature has been kind enough to humans by providing a wide range of plants having therapeutic potential. Screening of plants for isolation and identification of the natural bioactive products not only enrich the therapeutic compendium but also provide a cheaper, effective and safe alternative approach for treating diseases. It is a combined effort of botanists and clinicians for utilizing these plants for research and developing new drugs in controlling the growing epidemic of dreadful diseases such as myocardial infarction, diabetes, cancer, stroke etc. *Bombax ceiba* Linn. is one such plant species which possesses immense potential pharmacological actions. Last decade has revolutionized research on this plant showing its hypotensive, hypoglycemic, hypolipidemic, hepatoprotective, chemopreventive, antioxidant and fibrinolysis enhancing properties in various animal and human studies. This has led us towards the present compilation in order to provide a desk ready reference on everything which one needs to know about *B. ceiba*.

The present book entitled "Pharmacology of *Bombax ceiba* Linn." is a first monograph on the plant *Bombax ceiba* Linn., popularly known as Red Silk Cotton tree. It is in fact a compendium on this plant species containing total seven chapters and compiling all about *B. ceiba* starting from its historical and spiritual importance, distribution, botanical characterization and ethnobiological uses to modern phyto-pharmacy.

Bombax ceiba is well mentioned in the oldest written scriptures such as *Rigveda*, *Mahabharata* and *Ayurveda* along with a special role in many tribal cultures world wide. This handsome, deciduous tree is a part of tropical and subtropical forest ecosystem. Each part of the plant possesses immense medicinal value. A variety of chemical constituents including flavanoids, sesquiterpenoids, napthoquinones, phenols, steroids, carbohydrates and amino acids have been isolated and many are yet to be discovered. List of the phytoconstituents along with chemical structures of some important bioactive molecules is also provided in Chap. 3 for better understanding the chemistry of action. The fragmented research work done on its various biological activities in animal and human studies has been compiled in Chap. 4 with appropriate discussion. Further, toxicological profile of the plant is also discussed. Thorough analysis of its phytochemical profile and scientifically validated phyto-pharmacological properties will lead the way for research of newer phyto-pharmaceutical molecules beneficial for human health. Therefore, this book will prove a good reference material to go through.

The plant is not only rich in its history, ethnobiology, phytochemistry and pharmacology, but also possesses immense commercial and ecological importance. Hence, special chapters on these issues are provided which will be of interest to ecologists, agriculturists, foresters, industrialists and even to common people. Furthermore, the book has also incorporated some conservation strategies along with a case report on the sustainable conservation efforts done to preserve this plant species which is rapidly declining in many parts of the world due to ignorance of its importance. All photographs incorporated in the book are original and captured by authors themselves.

It is hoped that because of its vast scope and multifaceted coverage, the book will further accelerate the speed of research on this plant in various spheres world over which will rejuvenate this plant species for the betterment of health of future generation, a tree of infernal region will become the ornamental tree of gardens, an important project of research laboratories and a source of novel phyto-pharmaceutical compound for the treatment of dreadful diseases.

Undoubtedly, there may remain some errors in the text due to ignorance or misinterpretation of some aspects of the scientific literature. Authors welcome all critical comments and suggestions in order to improve the quality of future editions.

Acknowledgments

Authors thank the publishing house and all promoters of this project for their cooperation, support and for keen interest to bring this book in time. Thanks are also due to Dr. Preetesh Jain, USA, Prof. S.S. Katewa, Department of Botany, Mohanlal Sukhadia University, Udaipur and Dr. Phool Chander, Govt. Ayurvedic Officer, Punjab for providing immense help in literature collection. We also thank Mr. Rajesh Sharma for preparing an excellent artwork.

Dr. Vartika Jain
Dr. Surendra K. Verma

Contents

Abbreviations

ABTS	2,2′-Azino-di-[3-ethylbenzthiazoline sulphonate]
ACE	Angiotensin converting enzyme
ALT	Alanine transaminases
AST	Aspartate transaminases
ALP	Alkaline phosphatase
BMI	Body mass index
BP	Blood pressure
CCL_4	Carbon tetrachloride
COX	Cyclooxygenase
DMSO	Dimethyl sulfoxide
DNA	De-oxyribonucleic acid
DPPH	1,1-Diphenyl-2-picryl-hydrazyl
DW	Dry weight
EC	Effective concentration
ED	Effective dose
FA	Fibrinolytic activity
FAS	Fatty acid synthase
FRAP	Ferric reducing ability of plasma
g	Gram
GAE	Gallic acid equivalent
GPX	Glutathione peroxidase
GSH	Reduced glutathione
H_2O_2	Hydrogen peroxide
HAEC	Human aortic endothelial cell
HDL-C	High density lipoprotein cholesterol
HMG CoA	Hydroxymethylglutaryl coenzyme A
HUVEC	Human umbilical venous endothelial cells
IC	Inhibitory concentration
IL	Interleukin
iNOS	Inducible nitric oxide synthase
IgE	Immunoglobin E

I.P.	Intraperitoneal
I.V.	Intravenous
Kg	Kilogram
LD	Lethal dose
LDL-C	Low density lipoprotein cholesterol
LPS	Lipopolysaccharide
MIC	Minimum inhibitory concentration
n	Number of subjects
NO	Nitric oxide
NS	Not significant
OHA	Oral hypoglycemic agents
ORAC	Oxygen radical absorbance capacity
PGE	Prostaglandin E
SRB	Sulphorhoadmine B
TAS	Total antioxidant status
TBARS	Thiobarbituaric acid reactive substance
TEAC	Trolox equivalent antioxidant capacity
TLC	Thin layer chromatography
TNF	Tumar necrosis factor
TPA	12-O-tetradecanoylphorobol-13-acetate
UV	Ultraviolet
VLDL-C	Very low density lipoprotein cholesterol

Chapter 1
Introduction

Abstract *Bombax ceiba* Linn., a tree of ubiquitous occurrence, is characterized morphologically with sharp thorns, polyadelphous stamens and deciduous calyx; while anatomically it can be identified with concentric fibrous patches and rosette crystals of calcium oxalate. It is highly reputed in various traditional medicinal systems such as Ayurveda, Unani, Siddha and Traditional Chinese and Tibetan Medicine and considered as very good wound healer, tissue and bone regenerator, bowel-controller, anti-diarrheal, aphrodisiac, styptic agent and also useful in diseases such as kidney stones, burns and hyperpigmentation.

Keywords Bombacaceae · Silk-cotton tree · Polyadelphy · Microvita · Semal · Aphrodisiac · Diarrhea

The family Bombacaceae contains about 250 plant species grouped in approximately 30 genera. All species are arborescent, principally tropical and include some of the largest trees of the world. The genus *Bombax* includes 60 tropical species. The generic name '*Bombax*' refers to the cotton obtained from the fruits. In ancient India, the type species *Bombax ceiba* was known by the generic name *Salmalia* which is derived from Sanskrit language and conjures up a world of tradition and poetry. Earlier it was kept in the family Malvaceae but due to some important morphological differences such as trunk armed with thorns, polyadelphous stamens and pithy or wooly pericarp of fruits, a new family was created and named as Bombacaceae. In old Indian scriptures, *B. ceiba* is said to be the tree under which *Pitamaha* (The great grandfather-God) rested after the creation of world (Nicolson 1979; Shetty and Singh 1988; Santapau 1996).

V. Jain and S. K. Verma, *Pharmacology of Bombax ceiba Linn.*,
SpringerBriefs in Pharmacology and Toxicology,
DOI: 10.1007/978-3-642-27904-1_1, © The Author(s) 2012

1.1 Distribution

Bombax ceiba Linn. (syn.:*Bombax malabaricum* DC.; *Salmalia malabarica* (DC.) Schott & Endl.; *Gossampinus malabarica* (DC.) Merr.) is a lofty, deciduous tree (Fig. 1.1) and found in Temperate Asia (China to Taiwan), Tropical Asia (India, Bhutan, Cambodia, Indonesia, Philippines, Srilanka, Thailand, Malaysia, Laos and Vietnam), Papua New Guinea, Africa (Egypt) and Australia (Queensland, Northern territory and Western Australia). It is known by different names in different languages, such as Red Silk-Cotton tree, Indian Kapok tree (English), Shalmali (Sanskrit), Semal, Simal (Hindi), Shimul (Bengali), Bombax de Malabar, Cotonnier Mapou (French), Mu Mien (Chinese), Algodoeiro domatto, Arvore de Panha (Portuguese), Arbol capoc (Spanish), Indischer Seidenwolbaum (German) and Ngui (Thai) etcetera (Wightman and Andrews 1989; Chadha 1972).

It is a strong light-demander and fairly drought-resistant. It grows well in mixed deciduous, mixed evergreen and alluvial-savannah type of forests and prefers deep sandy-loam and alluvial soil of valleys. It thrives well in places where rainfall is 75–460 cm or more and well distributed throughout the year with maximum shade temperature from 34 to 49°C and minimum from 3.5 to 17.5°C (Chadha 1972). After initial protection from fire during sapling and seedling stages, it resists fire easily due to thick bark.

1.2 Botanical Characterization and Pharmacognostical Details

Genus *Bombax* is a member of family Bombacaceae under the order malvales of subclass dillenidae in magnliopsida. It includes 60 tropical species, among which *B. ceiba* is the most important species having 72 sporophytic chromosomes (Baum and Oginuma 1994). The generic name *Bombax* is a greek word adopted by Romans as an exclamation of surprise. Initially Linnaeus used the word *Xylon* for this tree which is a variant spelling of classical *Bombyx* referred for things of silk or cotton which is true for its cottony seeds and that is why it is better known as Silk-Cotton Tree (Nicolson 1979).

Bombax ceiba is a large, deciduous tree with a height up to 40 m, buttressed at base and horizontally spreading branches and armed with characteristic conical spines all over the stem leading to its name *Kantakdruma* (thorny tree) in Sanskrit language. However, these thorns are deciduous after 5–7 years. Roots of young plants, known as *Semal-musli* or *Semarkanda*, are medicinally very important. *Semal* is also known in Sanskrit as *Panchparni* (five-leaved) for its glabrous, entire, acuminate, elliptic-obovate or lanceolate and long-petiolated, alternate, palmately compound and generally penta-foliate leaves. Leaflets are glabrous with a size of 5−23 × 1.5−90 cm and secondary petioles are 2 cm in length. Stipules are small and caducous.

Fig. 1.1 *Bombax ceiba* Linn.
in full bloom

In Sanskrit, *Raktapushpa* (red flowered) and *Nirgandhpushpi* (flowers without smell) are its other names for the numerous bright, attractive, crimson red but odorless, actinomorphic, bisexual and hypogynous flowers which are crowded at the end of leafless branches. However, occasional specimens of white, pale-yellow colored flowers separately or among the red colored flowers are also reported (Santapau 1996; Bachulkar 2010). Pedicel is very short. Thick calyx is irregularly 3–5 lobed or cup shaped 3 cm long, silky on the inside and smooth outside. The margin is cleft into 3–5 valvate fringes. Corolla has five twisted and thick petals with a length of 12 cm or more and breadth of 2–3 cm. Polyadelphous androecium has 5–6 staminal bundles, each having 10–15 reniform stamens, an important identifying feature of this plant. Style is shortly clavate and little longer than stamens while stigma is pentafid and ovary is syncarpous, pentalocular with axile placentation. Another very important character is woody, 5-valved, oblong-ovoid, minutely apiculate, 10–15 cm long, 3–5 cm thick, dehiscent capsule, brownish-black when ripe with deciduous calyx and many 6–9 mm. long, obovoid, smooth and oily black seeds, enveloped in dense white silky hairs inside, produced during February to April which is its flowering and fruiting season (Davis and Mariamma 1965; Shetty and Singh 1988; Santapau 1996). A gummy exudate is obtained from the stem bark, called as *Mocharas* which is light-yellow initially and gradually becomes dark-brown (Fig. 1.2 a–o).

Fig. 1.2 Various parts of *B. ceiba*. **a** Young plant. **b** Spiny stem (*Kantakdruma*). **c** Leaf (*Panchparni*). **d** Longitudinally split bud and whole bud. **e** Flower (*Raktapushpa*). **f** Gum (*Mochrasa*) exuding from stem. **g** Gum in shape of *Lord Ganesha*. **h** Gum in shape of human heart. **i** Flowering branch. **j** Branches with fruits. **k** Unripe fruits. **l** Dehiscing capsules. **m** Split capsules. **n** Silk cotton. **o** Seeds

Fig. 1.2 (continued)

Fig. 1.2 (continued)

Macro and microscopic characters of root and stem of *B. ceiba* were typical of a dicotyledonous plant with distinguishing characters, such as presence of concentric series of fibrous patches alternating with groups of sieve elements in the secondary

Table 1.1 Analysis of stem bark of *Bombax ceiba* Linn. (The Ayuervedic Pharmacopoeia of India 2001)

Tests	Limits
Foreign matter	Not more than 1%
Total ash	Not more than 13%
Acid-insoluble ash	Not more than 2%
Alcohol-soluble extractive	Not more than 2%
Water-soluble extractive	Not more than 7%
Constituents	Saponins, tannins and gums
TLC of alcoholic extract	One fluorescent zone at Rf 0.59 (blue)
(solvent system: Toluene:	Three violet spots at Rf 0.44, 0.59 and 0.92
Ethyl acetate: 9:1)	(after spray with Vanillin-Sulfuric acid reagent)

Organoleptic characters: reddish-brown in color

phloem and presence of mucilage canals, tannin cells and cells containing rosette crystals of calcium oxalate in the ground tissue of root and stem (Mehra and Karnik 1968).

Macroscopically stem bark was found to be 0.5–1 cm thick, pale-ashy to silvery-gray externally, brownish internally, rough externally with vertical and transverse cracks, and mucilaginous on chewing. Powder of the stem bark is reddish-brown with fragments of cork cells, parenchymatous cells, single or groups of thick-walled, oval to irregular, stone cells having striations with narrow lumen (13–33 μ in diameter), rosette crystals of calcium oxalate, phloem fibers and numerous reddish-brown colored masses and tannin cells. Microscopically stem bark shows 10–15 layered, transversely elongated, radially arranged, thin-walled, cork cells with a few outer layers having brown-colored contents; thick-walled, oval to irregular stone cells in singles or in groups throughout the secondary cortex and concentric bands of narrow lumen and pointed tipped lignified fibers with numerous wavy phloem rays (The Ayurvedic Pharmacopoeia of India 2001).

Bagchi et al. (1992) have also demonstrated presence of acicular-shaped raphides (calcium oxalate crystals) distributed throughout secondary cortex, phloem parenchyma and ray cells of the stem bark. Mucilage canals and tannin cells were also found to be present in the parenchymatous cells of cortex. Table 1.1 describes some characters for determination of identity, purity and strength of stem bark of *B. ceiba* (The Ayurvedic Pharmacopoeia of India 2001).

1.3 Importance in Traditional Medicinal Systems

Besides various ethno-medicinal uses of *B. ceiba*, it is also well mentioned in indigenous systems of medicine such as Ayurveda, Unani, Siddha and Traditional Chinese and Tibetan medicine (Agnivesha-Charak-Dridhabala 2000; Liu 2009; Kabir 2002; Gyatso and Hakim 2010). Ayurveda, the Traditional Indian Medicine

Table 1.2 Ayurvedic standards of *Shalmali* (Bhavmishra 2010; Sharma 2001)

Parameter	Standard
Rasa (taste)	Madhur (sweet), Kashaya (astringent)
Guna (properties)	Laghu (light), Snigdha (unctuous), Picchila (sticky)
Virya (potency)	Sheeta (cooling)
Vipaka (post-digestive effect)	Madhura (sweet)
Karma (action)	Shothahara (anti-inflammatory), Kaphavardhaka (expectorant), Vedanasthapana (analgesic), Dahaprashamana (refrigerant), Grahi (astringent), Vrishya (virility promoter), Rasayani (lymphatics), Kashahar (thirst controller), Raktarodha (hemostatic)

describes the quality of plants by combining both the pharmcognosy (properties) and pharmacology (action). These traditional parameters reflect not only the quality but also efficacy of the plants (Table 1.2). Some of its medicinal uses and formulations as mentioned in Ayurveda are being described below in few paragraphs:

Powder of root (*Semal-musli*) with sugar is considered to be a good aphrodisiac. Root is also considered to possess anti-aging, anabolic and nutritive properties. Paste of leaves is applied in arthritis and on glandular swellings. Flowers of *B. ceiba* with seeds of *Papaver somniferum*, sugar and milk is prescribed three times a day to cure piles. Gum (*Mochrasa*) of the plant is styptic, analgesic, astringent, virility promoter and considered to be useful in menorrhagia/metrorrhagia, diarrhea and dysentery (Bhavmishra 2010; Agnivesha-Charak-Dridhabala 2002). Stem bark is said to be useful in hemorrhagic disorders, wound healing, removing pimples/acne and have a cooling effect in burning sensations. It is also used in hyperpigmentation, wounds, burns and stomatitis as a topical therapeutic agent (Gupta et al. 2004; Sushruta 2001).

Semal has been described among top ten drugs used as styptic, bowel regulator and tissue regenerator in Ayurveda. Pedicel/petiole of the plant or gum is used as enema in ulcerative colitis and dysentery (Agnivesha-Charak-Dridhabala 2000, 2002). *Mochrasa* of the plant is widely used in various Ayurvedic formulations for tissue regeneration, wound healing and anti-dysenteric effects (Vagbhatta 1993). Ayurveda describes the therapeutic doses of its various parts as follows: 5–10 g (Stem-bark, Root); 1–3 g (Fruit); 10–20 g (Flower) and 1–2 g of Gum (Agnivesha-Charak-Dridhabala 2000, 2002; The Ayurvedic Pharmacopoeia of India 2001).

A traditional formulation '*Shalmali ghrita*' prepared with flowers of *B. ceiba* is used as *Pramehagna* to cure polyurea, spermatorrhoea, leucorrhoea and menorrhagia (Sharma 2001). In painful micturition, a preparation called *Trinetra rasa* is given with a decoction in milk made of juice of *Cynodon dactylon*, liquorice root, gum of *B. malabaricum* and *Tribulus terrestris*. Prepared tin, mercury and sulfur is taken in equal parts, rubbed in an iron mortar and soaked seven times, respectively in the above-mentioned herb juices. Then it is roasted in covered crucible and again soaked in the above-mentioned fluid medicines and pills of four-grain are prepared. In diabetes, *Vanga bhasma* with honey, turmeric and juice of root of *B. malabaricum* is used (Nadkarni 1994). As a part of a traditional mixture, gum of the plant has also shown to cure giardiasis and has shown anti-motility, anti-diarrheal and anti-ulcer activities (Singh and Chaturvedi 1981; Bafna and Bodhankar 2003). Gum and flowers of *B. ceiba* have also been evaluated clinically in excessive bleeding per vaginum and showed significant decrease in blood loss along with improvement in menstrual cycle (Sinha et al. 2008).

In Unani medicine, seeds of *B. ceiba* are described as astringent and styptic, stem bark as diuretic, demulcent, tonic, aphrodisiac and used in uterus/vagina prolapse and impotency. Gum of the plant is a part of traditional Unani compound medicine *Mughalliz Mani* i.e. to make semen viscous; *Mumsik Mani* (Avaricious) i.e. to retain seminal discharge. Root powder is used in *Muallid Mani* to increase semen production and count (Kabir 2002). Flower powder has been shown to be effective in *Asrigdara* (dysfunctional uterine bleeding) and *Sailan-ur-Reham* (vaginal discharge) in various clinical studies as mentioned in Unani medicine (Prabha 1992; Fatima et al. 2000; Usmanghani 2011).

Bombax ceiba (*Muminahua*) is also an important ingredient of traditional Chinese formulations to strengthen spleen, promote digestion and eliminate food retention in intestines (Liu 2009). In traditional Tibetan medicine, flowers of *B. ceiba* are used to treat disorders of lung, heart and liver and pharmacologically found to possess cardiotonic, antibacterial and hepatoprotective activities (Gyatso and Hakim 2010). In Thai medicine, Jubliang is used for preparing herbal beverage. It is a mixture of eight herbs namely, *B. ceiba*, *Chrysanthemum morifolium*, *Imperata cylindrica*, *Lophatherum gracile*, *Nelumbo nucifera*, *Oroxylum indicum*, *Pragmites communis* and *Prunella vulgaris* and has shown to possess strong in vitro antioxidant and antimutagenic activity (Kruawan and Kangsadalampai 2006).

Many of these medicinal properties have been critically evaluated in various pharmacological studies proving thereby the ancient wisdom imparted to these indigenous medical systems.

References

Agnivesha-Charak-Dridhabala (2000) Charak Samhita Sutrasthana, chapter 4, Shloka 5, 31, 75, 46 and 77, 2nd edn. Chaukhamba Sanskrit Pratishthan, Delhi

Agnivesha-Charak-Dridhabala (2002) Charak Samhita Sidhisthana, chapter 10, Shloka 35. Chaukhamba Sanskrit Pratishthan, Delhi, p 967

Bachulkar M (2010) Note on a tree of *Bombax ceiba* L. with red and yellow flowers. Phytotaxonomy 10:143–146

Bafna P, Bodhankar S (2003) Gastrointestinal effects of Mebarid®, an ayurvedic formulation, in experimental animals. J Ethnopharmacol 86(2–3):173–176

Bagchi GD, Srivastava GN, Srivastava AK (1992) A study on the calcium oxalate crystals of some medicinal barks. Indian Drugs 29:561–567

Baum DA, Oginuma K (1994) A review of chromosome numbers in Bombacaceae with new counts for *Adansonia*. Taxon 43:11–20

Bhavmishra (2010) Bhavprakash chapter vatadivarga, Shloka 54-59. Chaukhamba Bharti Academy, Varanasi, pp 525–527

Chadha YR (1972) The Wealth of India, raw material, vol IX. Publications and Information Directorate, New Delhi

Davis TA, Mariamma KO (1965) The three kinds of stamens in *Bombax ceiba* L. (Bombacaceae) Bulletin du Jardin botanique de l'État a Bruxelles 35. Fasc.2

Fatima S, Ahmed J, Kabir H, Khan AJ (2000) Clinical evaluation of some Unani drugs *Sailan-ur-Reham Iltahabi* (inflammatory vaginal discharge). Hamdard Medicus 43(1):13–21

Gupta AK, Sharma M, Tandon N (2004) Reviews on Indian medicinal plants, Vol 4. Indian Council of Medical Research, New Delhi

Gyatso T, Hakim C (2010) Essentials of Tibetan traditional medicine. North Atlantic books, California

Kabir H (2002) Introduction to Ilmul Advia. Shamsher Publishers and Distributors, Aligarh

Kruawan K, Kangsadalampai K (2006) Antioxidant activity, phenolic compound contents and antimutagenic activity of some water extract of herbs. Thai J Pharm Sci 30:28–35

Liu Z (2009) Essentials of Chinese medicine, vol 3. Springer, Germany

Mehra PN, Karnik CR (1968) Pharmacognostic studies on *Bombax ceiba* Linn. Indian J Pharmacol 30:284

Nadkarni KM (1994) Indian material medica, vol 2. Popular Prakashan Ltd., Bombay

Nicolson DH (1979) Nomenclature of *Bombax ceiba* (Bombacaceae) and *Cochlospermum* (Cochlospermaceae) and their type species. Taxon 28(4):367–373

Prabha S (1992) Management and treatment of Asrigdar by *Shalmipushpa Churna*. In: Proceedings, 3rd international ayurvedic conference, Bali, Indonesia

Santapau H (1996) Chapter 7: silk-cotton tree. In: Santapau H (ed) Common trees. National Book Trust, New Delhi

Sharma P (2001) Dravyaguna Vijnanam, chapter 5. Chaukhamba Bharti Academy, Varanasi, pp 491–494

Shetty BV, Singh V (1988) Flora of India (Series 2), Flora of Rajasthan, vol 1. BSI publications, Calcutta

Singh KP, Chaturvedi GN (1981) Herbal treatment of Giardiasis. Sachitra Ayurveda 34(6):401–404

Sinha M, Shakuntala, Megha (2008) Clinical evaluation of *Shalmali* (*Salmalia malabarica*) in menorrhagia. J Res Educ Indian Med XIV(3):47–54

Sushruta (2001) Sushruta samhita sutrasthana chapter 38, Shloka 45-46, 12th edn. Chaukhamba Sanskrit Sansthan, Varanasi, pp 144–145

The Ayurvedic Pharmacopoeia of India (2001) Part I, Vol. III. Government of India, Department of Health and Family Welfare, Department of ISM & H, New Delhi. http://www.ayurveda.hu/api/API-Vol-3.pdf

Usmanghani K (2011) Unani medicine: implications and applications. http://botmatics.com/beta/demos/hamdard/papers-pdf/UNANI%20MEDICINE%20-Implications%20and%20applications.pdf. Accessed 18 Nov 2011

Vagbhatta (1993) Ashtang Hridya Sutrasthana chapter 15, Shloka 37-39, 11th edn. Chawkhamba Sanskrit Sansthan, Varanasi, p 107

Wightman G, Andrews M (1989) Plants of northern territory monsoon vine forests. Conservation Commission of the Northern Territory, Darwin

Chapter 2
Ethno-biology

Abstract Silk-cotton tree, at all the places of world where it is found in abundance, is an integral part of many socio-cultural religious ceremonies, rites, rituals, traditions, customs and festivals of native communities. It is an ancient tree species found in moist evergreen and deciduous forests of many continents. Almost every part of this tree is employed as medicine to treat various ailments of both humans and animals. It is in fact a tree with occult power and immense therapeutic potential. Many magico-religious beliefs associated with this tree have in fact been proved as a boon in disguise in its preservation on this planet earth.

Keywords *Holi* · Loogaroo · God tree · Mu Mien · Hot Qi · Impotence · Santeria · Black magic · *Shalmali*

2.1 Introduction

Bombax ceiba Linn. is an ancient tree of infernal region and has gained a reputation of 'Occult tree' in various communities worldwide. The tree has been well mentioned in the oldest documented written scripture of '*Rigveda*' and also mentioned in *Mahabharata* and the holy *Guru Granth Sahib* (Online document 1 2007; Dixit 1991).In the Jain scripture '*Uttaraadhyayan Sutra*'; to describe the aspects of soul, the legendary *Shalmali* tree is mentioned to signify the hellish state of our worldly existence; full of pain and misery (Online document 2 2011).

One of the oldest trees of *B. ceiba* (Gao tree) is present at Mo Pagoda, Nghi Duong Hamlet, Ngu Phuc Commune, Vietnam. This tree is 727 years old and recently in March, 2011, it has been denoted as the first Vietnam Heritage Tree by local people and Vietnam Association for Conservation of Nature and Environment (VACNE), so that people can take special care of it. This tree was planted by

Princess Quynh Tran in 1284 at Mo Pagoda where she used to go to worship Lord Buddha. She planted the tree to get the immense shadow provided by the tree where villagers and farmers can come and take rest for some time (Sinh 2011). Even the beautiful crimson-red flowers of the tree have been characterized as a flower emblem of Kaohsiung and Guangzhou cities, China (Online document 1 2007).

It is a part of food, medicine, folktales, customs and festivals in various indigenous communities of the world. As a food, tuberous tap roots and tender bark of young plants of *B. ceiba* are eaten during scarcity and otherwise also (Chadha 1972; Brock 1988). Tender leaves, flower buds, fleshy calyx and gum is also eaten raw (Arora 1997; Santapau 1996) or as a cooked vegetable (Angami et al. 2006). Flowers are made into a conserve by boiling with poppy seeds and sugar in goat's milk. Dried and powdered flowers are made into bread with or without corn (Chadha 1972) and dried stamens are also used as vegetable and as a curry powder in Thailand (Online document 3 2011). This chapter will deal with its ethno-biological importance particularly its medicinal uses, socio-cultural importance and associated magico-religious beliefs.

2.2 Medicinal Uses

Bombax ceiba has been deeply rooted in the minds of indigenous populations all over India and natives of South-East Asia. The plant is very popular for its use in seminal weakness in man and genital infections and inflammation in females. It is an important part of traditional five flowered herbal tree used to alleviate symptoms of 'Hot Qi' in Hongkong (Kong et al. 2006). Root, stem xylem and bark of *B. ceiba* with other herbs are important constituents of a Taiwanese folk medicine 'Mu-Mien' which is used as an anti-inflammatory and analgesic for the treatment of diarrhea, dysentery, hepatitis, liver cirrhosis, contused wound and scrofulosis (Lin et al. 1992). *Garo* tribe inhabiting North-East India; uses it in conjunction with other plants to treat jaundice (Rao and Negi 1980). Almost all parts of the plant are used for the management of general debility, weight loss, spermatorrhoea, premature ejaculation and sexual weakness in man and leucorrhoea and menstrual disorders in female. However, a wide range of ethnomedicinal uses has been described in the literature which is summarized in Table 2.1. Some of these ethno-medicinal recommendations have paved a way for development of herbal pharmaceutical combinations (Online document 5 2011) and still waiting for many more to come.

2.3 Socio-Cultural Importance

Plants are an integral part of human life and culture. *Bombax ceiba* is one such plant species which is also a part of many social and cultural events of many indigenous communities all over the world. There are many myths, legends, folktales, songs, customs and traditions associated with this large deciduous tree.

Table 2.1 Ethnomedicinal uses of various parts of *Bombax ceiba* Linn.

Plant part	Diseases	References
Root and bark	Diarrhea, dysentery, boils, burns	Jain (1991)
	Diabetes	Sarkar (1986), Maheshwari (2000), Rahman et al. (2008)
	Impotence	Sarkar (1986), Bhogaonkar and Kadam (2006)
	Aphrodisiac, Scorpion bite, Sex tonic	Oudhia (2003), Aswal (1996)
	Snake bite	Singh and Pandey (1998)
	Hyperpyrexia	Behera et al. (2006)
	Guinea Worm	Joshi (1989)
	Liver complaints	Jain (1989)
	Urinary troubles	Baruah and Sarma (1984), Mahato (2005), Guiley (2006)
	Common cold	Namhata and Mukarjee (1989)
	Brain tonic	Saghir et al. (2001)
	Gonorrhea	Dixit (1991), Singh et al. (1994)
	Syphilis	Chandra (1995)
	Bed Wetting, Leucorrhoea	Ghosh and Sensarma (1997)
	Spermatorrhoea	Aminuddin and Girach (1991)
Stem and bark	Dysentery due to bacterial, viral, protozoal or digestive disturbance	Reddy and Vatsavaya (2000)
	Boil	Shome et al. (1996)
	Heartburn	Singh and Pandey (1998)
	Heart tonic	Oudhia (2003)
	Leucorrhoea	Nadankunjidan (2005)
	Kidney stone	Katewa and Jain (2006)
	Spermatorrhoea, Weakness	Khandelwal (1997)
	Headache	Girach (1995)
	Acne	Saxena and Vyas (1981)
	Diabetes	Behera et al. (2006)
	Infertility	Guiley (2006)
	Blood vomiting	Kulip (2003)
	Body wash and swellings	Franco and Narsimhan (2009)
	Snake bite, scorpion, centipede and spider stings	Bakshi et al. (1999)
	Dislocated bones of animals	Sikarwar (1996)
	Conjunctivitis	Jain et al. (2010)
	Easy delivery of animals	Borthakur et al. (1996)
Gum	Asthma	Jain (1991)
	Giardiasis	Oudhia (2003)
	Bleeding Piles	Saradeshmukh and Nadkarni (1993)
	Diarrhea, dysentery	Singh and Pandey (1998)
	Menorrhagia and Leukorrhea	Singh et al. (1994)
	Scabies	Bakshi et al. (1999)
	Gonorrhea	Caius (2003)
	Kidney troubles and tuberculosis	Jain et al. (2010)
	Dental Caries	Chetty and Rao (1989)

(continued)

Table 2.1 (continued)

Plant part	Diseases	References
Leaf	Glandular swellings	Dixit (1991)
	Rheumatism	Chelvan (1998)
	Antidysenteric, Haematinic	Pandey et al. (1998)
	Anemia	Bakshi et al. (1999), Guiley (2006)
	Body pain	Samuel et al. (2010)
	Snake bite	Silja et al. (2008)
	Menorrhagia, Leucorrhoea	Tiwari and Majumdar (1996)
Flower	Hematuria	Khandelwal (1997)
	Anemia	Jain (1991)
	Leucorrhoea	Saini (1996), Khandelwal (1997)
	Hemorrhoids	Dixit (1991)
	Sexual diseases: Hydrocoele, Gonorrhea, Menstrual disorders, Leucorrhoea	Jain et al. (2004)
	Boils and sores	Manishi et al. (2005)
	Complete female sterilization	Tiwari et al. (1982)
	Menorrhagia	Negi et al. (1993)
	Splenomegaly, Internal bleeding, Cancer	Oudhia (2003)
	Colitis	Badhe and Pandey (1990)
	Premature ejaculation	Maheshwari et al. (1986)
	Snake bite	Tiwari and Majumdar (1996)
	Permanent Sterilization	Vedavathy et al. (1991), Bhuyan (1994)
	Diuretic, laxative	Shah and Gopal (1982)
	Stomach pain and Chicken pox	Varghese (1996)
	Cancer	Online document 4 (2006)
	Conjunctivitis	Jain et al. (2010)
	Urinary complaints of animals	Katewa and Jain (2006)
Fruit	Anti-fertility agent	Singh and Ali (1996)
	Uterus protrusion	Khandelwal (1997)
	Leucorrhoea	Tiwari and Majumdar (1996)
Fruit and Heartwood	Antidiabetic	Khan and Singh (1996)
	Antidiarrheal	Pandey et al. (1998)
	Snake bite	Badhe and Pandey (1990)
Seed	Chicken Pox, Small pox	Saxena et al. (1981)
	Abortion	Mitra and Mukherjee (2009)
	Spermatorrhoea	Khan and Khan (2005)
	Dysentery	Kunwar et al. (2009)
Cotton	Fresh burns and boils	Oudhia (2003)
	Piles	Sarkar (1986)
Thorn	Skin troubles, Acne	Sarkar (1986)
	Headache	Bakshi et al. (1999)

Due to the presence of sharp, conical thorns all over the trunk it is called as '*Kantakdruma*' in Sanskrit language. These thorns protect the plant from grazing animals and easy debarking. However, tribal communities of Rajasthan use this

Fig. 2.1 Tug-of-war
between stamens

tree as a shelter for young children of hens. They keep the animals in a natural sitting device made up of Bamboo mostly; known as '*Koiza*' and hang it on the top or in middle of the *Semal* tree. Due to the sharp pointed thorns, cats cannot climb on the tree and hence, the animals are easily saved (Sharma 2007).

It is also a part of natural masticator used by aboriginals. Sharp conical thorns of the tree are made blunt with stone and chewed fondly by tribal communities of Rajasthan. Its taste is said to resemble the nuts of *Areca catechu* (Joshi 1995). They also chew the thorns along with stem bark of *Cordia gharaf* to redden their lips and tongue at the occasion of festival and marriage (Singh and Pandey 1998).

Stamens of the plant are used to play with and to have fun in tribal communities in Rajasthan, India. The game is known as 'Tug of war'. Out of its numerous polyadelphous stamens, one group of stamen is selected and the opposed reniform anthers are interlocked in one another and stamen is pulled from both the players (Fig. 2.1). One who is left with the intact filament and anther is chosen as winner (Joshi 1995).

In ancient Bengal, India, soft Silk-cotton obtained from the fruits of *Semal* tree was used to prepare the dazzling red *Saris* (6 m long piece of cloth wore by Indian women) for young brides to wear at the time of their wedding. These *Saris* were not so durable but used to make bride very attractive during the wedding (Sarkar 1991).

Many multipurpose plant species become very popular among tribal communities either due to their material benefits, recreational opportunities or simply due to their beauty. These popular plant species are commonly sung by tribal communities in the form of folk songs. *Bombax ceiba* is also one such popular multipurpose tree species, merged in life and culture of tribal communities so much so that they identify themselves with *Semal*. In Udaipur, India, one of the major tribe known as '*Garasia*' sing a song "*Hemlo ropalo re*" in Mewari language meaning "O plant the hemlo" (*Semal* i.e. *B. ceiba*). In this song, moon and clouds has been given the status of its father and mother respectively and generally village-chief and his wife are assigned the role of its brother and sister-in-law and then a request is made to plant this tree and take care of it by considering it as one's own relative. A song is also sung to praise the dense shade of this large tree by *Garasia* tribe as they enjoy its shade very much (Joshi 1995).

During the months of March–April, when the tree is almost leafless yet embellished with bright red flowers, a *'Neja'* competition is organized by the tribal communities of Rajasthan, India. A bamboo basket is hung on one longest and largest *Semal* tree of the forest. Now the people are encouraged to procure the basket and whosoever does it in less time is declared as winner. However, this is in fact a blood shedding competition as the sharp thorns on its trunk do not make the way to basket an easy one (Jain 2009).

Bombax ceiba is an important part of festival *Holika-dahan/Holi* of India. *Holi* is known as 'festival of colors' and celebrated every year in spring season, when people enjoy harvesting their crops. It is that time of year when there is natural fragrance and color spread in the environment due to blooming of many flora. It is also blooming time of *B. ceiba* tree and its crimson-red flowers are also used to prepare eco-friendly color to play with. However, the preceding day of *Holi* festival is celebrated as *Holika-dahan* where a pole of debarked stem or branch of *B. ceiba* is used as a *Holi-danda* and burnt. This tradition is mainly present in states of northern India such as Rajasthan, Madhya Pradesh and Uttar Pradesh, where the tree is found abundantly (Jain et al. 2009). However, from East to West, the use of sacred cotton tree (*Semal*) in *Holi* is a must in India (Crooke 1914).

Bhil, *Garasia* and *Damor* tribes residing particularly in southern Rajasthan consider this tree as a mythological character *'Prahlad'*, who was an ardent devotee of Lord Vishnu, and survived in fire in spite of a killing attempt made by his wicked aunt *'Holika'* as per orders of his father King *Hirankashipu*. This is enacted every year where the debarked stem of *Semal* is covered with dry grass and erected for burning. The burning of *Holika* is celebrated as *Holi* or *Holika-Dahan*. The same scene is enacted every year to remind people that those who love God shall be saved, and those who torture the devotees of God shall be reduced to ashes. The concept behind using this particular tree as *Prahlad* might be its fire resistant property due to which it does not burn so easily and thus survive for a long period in fire as *Prahlad* survived in the tale (Jain et al. 2009).

However, there also exist some different trends in different tribes residing in different districts of South Rajasthan, India but use of *Semal* tree is a must. In *Bhil* tribe, before cutting a *B. ceiba* pole, a coconut is tied on the bough, liquor trickled and vermillion is applied and tree is cut in such manner so as to have a head and two arms and generally the pole is removed from the burning pile. This traditional two armed *Holi* is an important part of Udaipur, India (Figs. 2.2a, b).

In some *Bhil* villages of Banswara, India; Bamboo representing as *Prahlad* is tied in a red cloth and planted along with *Semal* which is considered as the wicked aunt *Holika*. *Kathodi* tribe residing in Udaipur, plant five poles of five different plant species for *Holika-dahan* and *Semal* is also one of them. They keep length of each pole to the level of a person's head and before planting the pole, a coin and *Areca catechu* nut is placed at the spot and all the poles are allowed to be burnt

Fig. 2.2 Traditionally decorated two armed *Holi*. **a** Rural *Holi;* **b** Royal *Holi*

(Joshi 1995). However, this tradition has made a great impact on eco-system, particularly in Udaipur district with a significant reduction in number of *Semal* trees (Jain et al. 2007) and has become a point of discussion recently (Jain et al. 2009).

2.4 Magico-Religious Beliefs

These beliefs are associated with auspiciousness or bad luck causing elements present among the indigenous communities. *Bombax ceiba* is a part of many mythological stories of traditional folk such as Indian, Caribbean, French and African population. These all reflect its ubiquitous occurrence and deep rooted magico-religious beliefs of the people. Some of these are indirect projections of mentality showing concern for its sustainable use and conservation while some are seriously damaging its existence in nature.

The two contrasting beliefs come across while searching literature on *B. ceiba*. One thought considers it to be a tree of Devils. On the other hand, it has also got reputation of spiritual sacred tree; known as '*Dev Vriksa*' in Sanskrit language and the whole plant is worshipped (Jain and Borthakur 1980). Both the concepts are quite well prevalent throughout world in many indigenous communities.

It is told to be a tree of infernal region. In Sanskrit language, the tree is also called as '*Yamadruma*' which means that it is 'tree of *Yama*'-the lord of death. The thorny appearance of the tree resembles the weapon of *Yama* as depicted in the mythological scriptures and thus people are afraid of cutting this tree

(Warrier et al. 1994). The tree is also known as Devil's tree/Jumbies tree in Grenada, Caribbean islands. Silk-cotton tree is considered to be a place where Vampires, known as Loogaroo keep their skins safely. Loogaroo are generally old women who make a pact with devil to learn some magic and in turn provide them fresh blood. They roam in the city in search of fresh blood in the night and as the morning starts, they again go to the tree and wear their skins back (Summer 2003).

Nagualism was a mysterious and powerful cult practiced by Mexican and Central American tribes. The nagualistic rites were highly symbolic, and the symbols had clearly defined meanings. One of the symbol still venerated as a survival of ancient cult is that of a tree species— *B. ceiba*. The conventionalized form of this tree strongly resembles a cross, and this came to be the ideogram of "life" (Brinton 1894).

Santeria is an animistic religion of West African and Caribbean origin, influenced by Roman Catholic Christianity. *Santeros* also consider *B. ceiba* tree as sacred and worship it as a female saint. Devotees do not cross the tree's shadow without permission. The tree helps them in casting of spells and to harm the victims. *Santeros* also perform black magic and to become masters (*Mayombero*) in this, a great effort is required where the supplicant must sleep under this tree for seven days and then takes a new suit of clothes and bury that in a previously selected grave. After becoming *Mayombero*, he receives crown of its leaves and prepare his own *nganga*, the cauldron with all his magical portions and powers (Guiley 2006).

The Luo are a sub group of Yi ethnic group residing in border areas of Yunnan, Guangxi and Vietnam. In this group, after death of a person, there is a tradition of sitting together and narration of many stories and myths revealing Luo values and traditional customs to the living audience. The Red Silk-Cotton tree is the name of a folktale of Luo group which starts with this handsome tree and is an another example showing the immense importance of this tree among native communities (Mair and Bender 2011).

Another prevalent myth also infusing fear in the minds is that if a healthy person visualizes this tree even in dreams, he will become ill and if an ill person sees it in dreams; he will be dying soon (Mahatma 1968). In Dungarpur district of Udaipur division, India, it is considered as inauspicious, due to hooting of owls that make their home on it (Joshi 1995).

Bhil tribe inhabitant of South Rajasthan, India has another superstitious fear in their minds which is associated with the silk-cotton obtained from fruits of *B. ceiba*. According to them, mattresses and pillows filled with its plumed seeds will cause paralysis of the user and so they do not use its cotton (Singh and Pandey 1998). This myth is in fact protecting its natural dispersal source- the seeds, which are embedded in the silk-cotton. Wood of this tree is not used as a fuel source by the tribal communities of Rajasthan as they believe that it will bring bad luck to them (Arora 1997). One of the oldest scripture-*Brahamavaivarta Purana* prohibits use of this plant for brushing the teeth (Bakshi 1999). However, all these inauspicious tales associated with this plant has led indirectly for its survival and preservation in nature for a long time.

It is one of the preferred species for Honey bees for keeping their hives. Many bee hives can be seen on a single huge *Semal* tree. These hives are a good source of honey—a major non-wood forest product. Every year during summer season, tribal communities of southern Rajasthan perform a Goat sacrifice ceremony before removing the bee hives from the tree to please the honey bees so that they will again come on the same *Semal* tree next year. The meat of Goat is served as a *'Prasada'* (sacred food) among the tribal communities (Sharma 2007). It is one of the sacred trees to the Austro-Asiatic Godabas and Bondos of Orissa, India who celebrate megalithic ceremonies in front of this tree where the buffaloes and bovines are tied to the branches of the tree and sacrificed, menhirs erected and *Semal* trees are planted (Brighenti 2001).

Trees were considered as hypaethral i.e. roofless temples in Vedic religion. *Bombax ceiba* has also been considered as God tree since vedic times. The tree is mentioned in the oldest scriptures; *Rigveda, Mahabharata* and *Guru Granth Sahib* (Online document 1 2007; Dixit 1991). It is considered to be a part of *'Panchwati'* where five spiritual trees were planted together to attract positive microvita (Sarkar 1991) to help in practicing psycho-spiritual meditation in ancient times. *Vishnu Dharmottar Puran*, recommend *B. ceiba* for plantation in sacred groves. *Bhavishya Puran* describes that this tree should be ritually invoked before plantation (Agarwal 2010).

Vedic religion believes that every individual is born on certain constellation which is regarded as protector of that individual. These constellations are associated with specific trees called as *'Nakshatra trees'* which are worshipped and treated as equivalent to God. There are total 27 constellations and *B. ceiba* is said to be *nakshatra tree* of people born in 18th constellation—*'Jyestha'*. People borne in *'Jyestha'* constellation are advised to plant *B. ceiba* and they cannot use the tree for medicine (Sane and Ghate 2006).

Bombax ceiba is called to be home of *'Yakshis'*; the female tree spirits. These *Yakshis* are worshipped by women for gift of children and thus *Semal* tree is indirectly conserved for a long time (Gupta 1995). Sacred groves are groups of some auspicious plants located near the worship places of tribal communities. *Bombax ceiba* is also a part of sacred grove called as *'Maad Bavas*i' in Bosa village in Rajasthan, India. This is a worship place of *Garasia* tribe and thus no part of this tree is allowed to be cut or used for any purpose to avert the wrath of *Maad Bavasi*. Hence, sacred grooves are important places helping in conservation of plant species. Not only this, but the beautiful, odorless bright red and large flowers of *Semal* are also offered to local god and goddesses (deities) by tribal communities and thus the tree is considered as sacred one. *Semal* is also considered as a tree totem by *'Semlia'* clan of *Bhil* tribe dwelling in Rajasthan and hence respected, worshipped and protected (Joshi 1995).

2.5 Ethno-conservation

Many plants are part of ethno-conservation practices and *B. ceiba* is not an exception. Community-based ethno-conservation mechanism of plants is prevalent among *Meetei* community in Manipur, India. Khuman clans of *Meetei* community do not use or consume *B. ceiba* in any form and thus by employing simple environmental ethics in consumption or harvesting, conserve the tree (Singh et al. 2006).

In the states of Madhya Pradesh and Chattisgarh, India, tribal communities use various strategies to protect *Semal* tree using their environment friendly ethno-conservation approaches. For example, usually Mondays are preferred days to collect roots of the plant and mostly roots spreading in north direction are collected and in this way they balance the collection frequency and thus avoid unnecessary harm and exploitation of the plant. Moreover, root, leaves and other plant parts are not collected simultaneously and for collection of every part, a specific worship procedure is performed. The most useful strategy for medicine collection includes clock wise selection of different patch in a group of *Semal* trees which helps in sustainable conservation. These rules are stringent upon traders from other areas too (Oudhia 2003).

In a nutshell, wherever this tree species is present, it has affected the human sentiments attached to their magico-religious beliefs and influenced the socio-cultural customs. It has nourished human beings and animals by providing food; infused health and generated psycho-spiritual vibrations for mental power. In fact, it has proved itself as the benevolent colleague of humanity.

References

Agarwal S (2010) Daan and other giving traditions in India: The forgotten pot of gold. Account Aid, New Delhi

Aminuddin, Girach RD (1991) Ethnobotanical studies on Bondo tribe of district Koraput (Orissa), India. Ethnobotany 3:15–19

Angami A, Gajurel PK, Rethy P, Singh B, Kalita SK (2006) Status and potential of wild edible plants of Arunachal Pradesh. Indian J Trad Knowl 5(4):541–550

Arora A (1997) Ethnobotanical studies on the wild plants from Aravalli hills of Rajasthan. Ph.D. Thesis, Mohanlal Sukhadia University, Udaipur

Aswal BS (1996) Conservation of ethno-medicinal plants diversity of Garhwal Himalaya in India. In: Jain SK (ed) Ethnobiology in human welfare. Deep Publications, New Delhi

Badhe PD, Pandey VK (1990) A study of medicinal and economic plants of Amravati division, Amravati circle, Maharashtra. Bull Med Ethnobot Res 11:1–3

Bakshi DN, Sensarma P, Pal DC (1999) A lexicon of medicinal plants in India, Naya Prakash, Calcutta. vol I. pp 360–365

Baruah P, Sarma GC (1984) Studies on the medicinal uses of plants by the Boro tribals of Assam.II. J Econ Tax Bot 11:71–76

Behera SK, Panda A, Bhera SK, Misra MK (2006) Medicinal plants used by the Kandhas of Kandhamal district of Orissa. Indian J Trad Knowl 5(4):519–528

Bhogaonkar PY, Kadam VN (2006) Ethnopharmacology of Banjara tribe of Umarkhed taluka, district Yavatmal, Maharashtra for reproductive disorders. Indian J Trad Knowl 5(3):336–341

Bhuyan DK (1994) Herbal drugs used by the tribal people of Lohit district of Arunachal Pradesh for abortion and easy delivery-A report. Adv Plant Sci 7:197–202

Borthakur SK, Sarma VK (1996) Ethno-veterinary medicine with special reference to cattle prevalent among the Nepalis of Assam, India. In: Jain SK (ed) Ethnobiology in human welfare. Deep Publications, New Delhi

Brighenti F (2001) Sakti cult in Orissa. D.K Printworld, New Delhi

Brinton DG (1894) Nagualism: A study in native American folk-lore and history. McCalla, Philadelphia

Brock J (1988) Top end native plants. J Brock, Darwin

Caius F (2003) The medicinal and poisonous plants of India. Scientific Publishers, Jodhpur

Chadha YR (1972) The Wealth of India. Raw material. Vol IX, Publications and Information Directorate, New Delhi

Chandra K (1995) An ethnobotanical study on some medicinal plants of district Palamau (Bihar). Sachitra Ayurved 48:311–314

Chelvan PT (1998) Traditional phytotherapy among the migrant Tamilian settlers in South Gujarat. Biojournal 10:9–14

Chetty KM, Rao KN (1989) Ethnobotany of Sarakallu and adjacent areas of Chittoor district, Andhra Pradesh. Vegetos 2:51–58

Crooke W (1914) The Holi: A vernal festival of the Hindus. Folklore 25(1):55–83

Dixit I (1991) An ethnobotanical survey of ingredients of domestic remedies in use in Ajmer district (Rajasthan). Ph.D. Thesis, University of Rajasthan, Jaipur

Franco MF, Narsimhan D (2009) Plant names and uses as indicators of knowledge patterns. Indian J Trad Knowl 8(4):645–648

Ghosh S, Sensarma P (1997) Ethnomedicine to modern medicine: An observational study in some villages of West Bengal. Ethnobotany 9:80–84

Girach RD (1995) Ethnomedicinal uses of plants among the tribals of Singhbhum district, Bihar, India. Ethnobotany 7:103–107

Guiley R (2006) The encyclopedia of magic and alchemy. Infobase Publishing, New York

Gupta SM (1995) Woman and tree motifs, 2nd edn. In: Jain SK (ed) A manual of Ethnobotany. Scientific Publishers, Jodhpur

Jain A, Katewa SS, Chaudhary BL, Galav P (2004) Folk herbal medicines used in birth control and sexual diseases by tribals of Southern Rajasthan, India. J Ethnopharmacol 90:171–177

Jain DL, Baheti AM, Jain SR, Khandelwal KR (2010) Use of medicinal plants among tribes of Satpuda region of Dhule and Jalgaon districts of Maharashtra—an ethnobotanical survey. Indian J Trad Knowl 9(1):152–157

Jain SK (1991) Dictionary of Indian folk medicine and ethnobotany. Deep Publications, New Delhi

Jain SK, Borthakur SK (1980) Ethnobotany of Mikirs of India. Econ Bot 34:264–272

Jain SP (1989) Tribal remedies from Saranda forest, Bihar, India -I. Int J Crude Drug Res 27:29–32

Jain V (2009) Isolation of active principles and effect of crude drugs obtained from *Ipomoea digitata* Linn. and *Bombax ceiba* Linn. for their antioxidant property vis-à-vis endothelial dysfunction in human beings. Ph.D. Thesis, Mohanlal Sukhadia University, Rajasthan

Jain V, Verma SK, Katewa SS (2007) A dogmatic tradition posing threat to *Bombax ceiba*—the Indian Red Kapok tree. Medicinal Plant Conserv 13:12–15

Jain V, Verma SK, Katewa SS (2009) Myths, traditions and fate of multipurpose *Bombax ceiba* L.—an appraisal. Indian J Trad Knowl 8(4):638–644

Joshi P (1989) Herbal drugs in tribal Rajasthan from child birth to child care. Ethnobotany 1:77–87

Joshi P (1995) Ethnobotany of primitive tribes in Rajasthan. Printwell, Jaipur

Katewa SS, Jain A (2006) Traditional folk herbal medicines. Apex Publishing House, Udaipur

Khan MA, Singh VK (1996) A folklore survey of some plants of Bhopal district forests, Madhya Pradesh, India, described as antidiabetics. Fitoterapia 67:416–421

Khan VA, Khan AA (2005) Herbal folklores for male sexual disorders and debilities in Western Uttarpradesh. Indian J Trad Knowl 4(3):317–324

Khandelwal S (1997) Ethnobotany of the Bhil tribe in Rajasthan. Ph.D. Thesis, University of Rajasthan, Jaipur, India

Kong FY, Ng DK, Chan CH et al (2006) Parental use of the term "Hot Qi" to describe symptoms in their children in Hong Kong: a cross sectional survey "Hot Qi" in children. J Ethnobiol Ethnomedicine 5:2–8

Kulip J (2003) An ethnobotanical survey of medicinal and other useful plants of Muruts in Sabah Malaysia. Telopea 10(1):81–98

Kunwar RP, Uprety Y, Burlakoti C, Chowdhary CL, Bussmann RW (2009) Indigenous use and ethnopharmacology of medicinal plants in far-west Nepal. Ethnobotany Res Appl 7:5–28

Lin CC, Chen SY, Lin JM, Chiu HF (1992) The pharmacological and pathological studies on Taiwan folk medicine (VIII): The anti-inflammatory and liver protective effects of "Mu-mien". Am J Chin Med 20(2):135–146

Mahatma U (1968) Dhanvantri, Vanoushdhi Visheshank (VI part). Choukhambha Prakashan, Varanasi

Mahato RB, Chaudhary RP (2005) Ethnomedicinal plants of Palpa district, Nepal. Ethnobotany 17:152–163

Maheshwari JK (2000) Ethnobotany and medicinal plants of Indian subcontinent. Deep Publications, New Delhi

Maheshwari JK, Kalakoti BS, Lal B (1986) Ethnomedicine of Bhil tribe of Jhabua district, Madhya Pradesh. Anc Sci Life 5:255–261

Mair VH, Bender M (2011) The Columbia anthology of Chinese folk and popular literature. Columbia University Press, NewYork

Manishi P, Srinivasa BH, Shivanna MB (2005) Medicinal plant wealth of local communities in some villages in Shimoga district of Karnataka. J Ethnopharmacol 98:307–312

Mitra S, Mukherjee SK (2009) Some abortifacient plants used by the tribal people of West Bengal. Nat Prod Radiance 8(2):167–171

Nadankunjidan S, Abirami S (2005) Comparative study of traditional medical knowledge of Pondicherry and Karaikal regions in union territory of Pondicherry. Ethnobotany 17:112–117

Namhata D, Mukarjee A (1989) Some common practices of herbal medicines in Bankura district, West Bengal. Indian J Forestry 12:318–321

Negi KS, Tiwari JK, Gaur KD, Pant KC (1993) Notes on ethnobotany of five districts of Garhwal Himalayas, India. Ethnobotany 5:73–81

Online document 1. *Bombax ceiba*. http://en.wikipedia.org/wiki/Bombax_ceiba. Accessed 17 Feb 2007

Online document 2. Concept of soul in Jain scriptures. Quotations from Uttaraadhyayan Sutra. http://www.jainstudy.org/jsc4.00-QfromS-ConceptOSoul.htm. Accessed 25 Nov 2011

Online document 3. Local vegetables of Thailand. JIRCAS. http://www.jircas.affrc.go.jp/project/value_addition/Vegetables/020.html. Accessed 28 Jun 2011

Online document 4. Oudhia P. Traditional medicinal knowledge about herbs used in treatment of cancer in Chhattisgarh, India. XXVI. Interactions with the traditional healers of Dhamtari region. http://ecoport.org. Accessed 30 Oct 2006

Online document 5. www.favorfinesse.com/hhacnecream.shtml. Accessed 25 Jan 2011

Oudhia P (2003) Articles by Pankaj Oudhia. http://www.botanical.com/site/column_poudhia/151_semai.html. Accessed 21 Jan 2011

Pandey BN, Das PKL, Jha AK, Ojha AK, Mishra SK, Yadav S (1998) Ethnobotanical profile of South Bihar with special reference to East Singhbhum (Jamshedpur). Acta Ecol 20:31–38

Rahman AHMM, Anisuzzaman M, Haider SA, Ahmed F, Islam AKM, Naderuzzaman ATM (2008) Study of medicinal plants in the graveyards of Rajshahi city. Res J Agric Biol Sci 4(1):70–74

Rao RR, Negi B (1980) Observation on the ethnobotany of the Khasi and Garo tribes in Meghalaya. J Econ Tax Bot 1:157–162

Reddy S, Vatsavaya SR (2000) Folklore biomedicine for common veterinary diseases in Nalgonda district, Andhra Pradesh, India. Ethnobotany 12:113–117

Saghir IA, Awan AA, Majid S, Khan MA, Qureshi SJ, Bano S (2001) Ethnobotanical studies of Chikar and its allied areas of district Muzaffarabad. J Biol Sci 1(12):1165–1170

Saini VK (1996) Plants in the welfare of tribal women and children in certain areas of central India. In: Jain SK (ed) Ethnobio Human Welf. Deep Publications, New Delhi

Samuel AJSJ, Kalusalingam A, Chellappan DK, Gopinath R, Radhamani S, Husain HA, Muruganandham V, Promwichit P (2010) Ethnomedical survey of plants used by the Orang Asli in Kampung Bawong, Perak, West Malaysia. J Ethnobiol Ethnomedicine 6:5–10

Sane H, Ghate V (2006) Sacred conservation practices at species level through tree worship. Ethnobotany 18:46–52

Santapau H (1996) Chapter 7: Silk-cotton tree. In: Santapau H (ed) Common trees. National Book Trust, New Delhi

Sardeshmukh R, Nadkarni K (1993) Raktarsha and mocharasa yoga-a study. Deerghayu Int 11(2):14

Sarkar PR (1986) Yaogic treatment and Natural remedies, 2nd edn. AMPS Publication, Calcutta

Sarkar PR (1991) Microvitum in a nutshell, 3rd edn. AMPS Publication, Calcutta

Saxena AP, Vyas KM (1981) Ethnobotanical records on infectious disease from tribals of Banda district (UP). J Econ Tax Bot 2:191–195

Saxena HO, Brahman M, Dutta PK (1981) Ethnobotanical studies in Orissa. In: Jain SK (ed) Glimpses of Indian ethnobotany. Deep Publications, New Delhi

Shah GL, Gopal GV (1982) An ethnobotanical profile of the Dangies. J Eco Tax Bot 3:355–364

Sharma SK (2007) Study of biodiversity and ethnobiology of Phulwari Wildlife Sanctuary, Udaipur (Rajasthan), Ph.D. Thesis, Mohanlal Sukhadia University, Udaipur

Shome U, Rawat AKS, Mehrotra S (1996) Time tested household herbal remedies. In: Jain SK (ed) Ethnobiol Human Welf. Deep Publications, New Delhi

Sikarwar RLS (1996) Ethno-veterinary herbal medicines in Morena district of M.P. India. In: Jain SK (ed) Ethnobiol Human Welf. Deep Publications, New Delhi

Silja VP, Varma KS, Mohanan KV (2008) Ethnomedicinal plant knowledge of the Mullu kuruma tribe of Wayanad district Kerala. Indian J Trad Knowl 7(4):604–612

Singh JL, Singh BN, Gupta AK (2006) Environmental ethics in the culture of Meeteis from North East India. www.eubios.info/ABC4/abc4320.htm. Accessed 25 Jun 2011

Singh KK, Kalakoti BS, Prakash A (1994) Traditional phytotherapy in the health care of Gond tribals of Sonbhadra district, UttarPradesh, India. J Bombay Nat Hist Soc 91:386–390

Singh V, Pandey RP (1998) Ethnobotany of Rajasthan. Scientific Publishers, Jodhpur

Singh VK, Ali ZA (1996) Folk medicinal plants used for family planning in India. In: Ethnobiology in human welfare. Jain SK (ed) Deep Publications, New Delhi

Sinh NN (2011) The first ancient "Gao" tree (*Bombax ceiba*) to be honored as the Vietnam Heritage Tree. http://vacne.org.vn/en/default.aspx?newsid=1044. Accessed 24 Jun 2011

Summer M (2003) The Vampire His kith and kin. Kessinger Publisher, MT

Tiwari KC, Majumdar R, Bhattacharjee S (1982) Folklore information from Assam for family planning and birth control. Int J Crude Drug Res 20(3):133–137

Tiwari KC, Majumdar R (1996) Medicinal plant from upper Assam borders having specific folk uses. Sachitra Ayurveda 49:207–215

Varghese E (1996) Applied ethnobotany—a case study among the Kharias of Central India. Deep Publications, New Delhi

Vedavathy S, Rao KN, Rajaiah M, Nagaraju N (1991) Folklore information from Rayalaseema region, Andhra Pradesh for family planning and birth control. Int J Pharmacognosy 29:113–116

Warrier PK, Nambiar VPK, Ramankutty C (1994) Indian medicinal plants—a compendium of 500 species, vol 1. Orient Longman Publishing, Kottakkal

Chapter 3
Phytochemical Studies

Abstract Many chemical compounds have been isolated from different parts of
B. ceiba worldwide. These belong mostly to phenolics, flavonoids, sesquiterpenoids, steroids, naphthoquinones and neolignans. Totally 16 compounds have been
isolated from root, 8 from root bark, 3 from stem bark, 3 from heart wood, 2 from
leaves, 78 from flowers, 19 from seeds and 11 from gum. Several of these compounds are novel and isolated for the first time. Compounds isolated from this
plant possess very important biological activities including immunomodulatory,
cytotoxic, antineoplastic, anti-inflammatory, hypotensive, hypolipidemic, antihyperglycemic and antioxidant.

Keywords Sesquiterpenoids · Napthoquinone · Phenolics · Lupeol · Mangiferin ·
Anti-HIV

3.1 Introduction

Chemical studies on *B. ceiba* were started in 1935 by T.P. Ghose when he
chemically analyzed its young roots for the first time. Since then, many compounds have been isolated from its various parts out of which the novel ones are
Shamimicin, Bombamalosides, Bombamalones, Bombasin, Bombasin 4-*O*-glu-
coside and Bombalin which have been isolated first time from any plant species.
A list of the compounds isolated from various parts of the plant have been presented in Table 3.1. Chemical structures of some of the compounds isolated from
B. ceiba are also provided in Fig. 3.1.

V. Jain and S. K. Verma, *Pharmacology of Bombax ceiba Linn.*,   25
SpringerBriefs in Pharmacology and Toxicology,
DOI: 10.1007/978-3-642-27904-1_3, © The Author(s) 2012

Table 3.1 Chemical compounds isolated from various parts of *Bombax ceiba*

Plant part	Compounds	References
Root	n-triacontanol	Chauhan et al. (1980)
	β-sitosterol	
	5,7,3,4-tetrahydroxy-6-methoxyflavan-3-O-β-D-glucopyranosyl-α-D-xylopyranoside	
	1,6-dihydroxy-3-methyl-5-(1-methylethyl)-7-methoxy-8-carboxylic acid (8→1 lactone), $C_{16}H_{16}O_4$ [1]	Sood et al. (1982)
	Daucosterol	Qi et al. (1996)
	Oleanolic acid	
	Hesperidin	
	Potassium nitrate	
	Isohemigossylic acid lactone 2-methyl ether [2]	Puckhaber and Stipanovic (2001)
	Bombamalone A: 3,7-dihydroxy-5-isopropyl-8-methoxy-7-methyl-7H-naphthol [1,8-bc] furan-2,6-dione, $C_{16}H_{16}O_6$	Zhang et al. (2007)
	Bombamalone B: 10-hydroxy-3,5,5,10-tetramethylbenzofuro [4,3-fg] chromene-1,2(5H, 10H)-dione, $C_{18}H_{16}O_5$	
	Bombamalone C: 7-hydroxy-4-isopropyl-8-methoxy-6-methyl naphthalenene-1,2-dione, $C_{15}H_{16}O_4$	
	Bombamalone D: 5,8-dihydro-2-hydroxy-4-isopropyl-7-methoxy-6-methyl-5,8-dioxonaphthalene-1-carboxylic acid, $C_{16}H_{16}O_6$	
	Bombamaloside:4-O-β-glucopyranosyl-6,7-dihydro-2,2,8-trimethyl-6,7-dioxo-2H-naphthol [1,8-bc] furan-5-carboxylic acid, $C_{21}H_{22}O_{11}$	
	Lacinilene	
	Bombaxquinone B	

(continued)

Table 3.1 (continued)

Plant part	Compounds	References
Root bark	Lupeol [3]	Sheshadri et al. (1971)
	β-sitosterol	
	Isohemigossypol-1-methyl ether	Sankaram et al. (1981)
	Isohemigossypol-1,2-dimethyl ether	
	8-formyl-7-hydroxy-5-isopropyl-2-methoxy-3-methyl-1, 4-naphthaquinone, $C_{16}H_{16}O_5$ [4]	
	7-hydroxycadalene, $C_{15}H_{18}O$ [5]	
	5-isopropyl-3-methyl-2,4,7 trimethoxy-8,1-naphthalene carbolactone, $C_{18}H_{20}O_5$ [6]	Reddy et al. (2003)
	8-formyl-7-hydroxy-5-isopropyl-2-methoxy-3-methyl-1, 4-naphthaquinone, $C_{16}H_{16}O_5$ [4]	
Stem bark	Lupeol, $C_{30}H_{50}O$ [3]	Mukherjee and Roy (1971)
	β-sitosterol, $C_{29}H_{50}O$	
	Shamimicin:1,1-bis-2- (3,4-dihydroxyphenyl)-3,4-dihydro-3,7-dihydroxy-5-O-xylopyranosyloxy-2H-1-benzopyran	Saleem et al. (2003)
	Lupeol, $C_{30}H_{50}O$ [3]	
Heart wood	7-hydroxy-5-isopropyl-2-methoxy-3-methyl-1,4-naphthoquinone, $C_{15}H_{17}O_4$ [7]	Sreeramulu et al. (2001)
	7-hydroxycadalene, $C_{15}H_{18}O$ [5]	
	8-formyl-7-dydroxy-5-isopropyl-2-methoxy-3-methyl-1,4-naphthoquinone, $C_{16}H_{16}O_5$ [4]	
Leaves	Shamimin: 2- (2,4,5-trihydroxy phenyl)-3,5,7-trihydroxy-6-C-glucopyranosyloxy-4H-1- benzopyran-4-one	Faizi and Ali (1999); Saleem et al. (1999)
	Mangiferin: 2-β-D-gluopyranosyl- 1,3,6,7-tetrahydroxy-9H-Xanthen-9-one [8]	Shahat et al. (2003); Dar et al. (2005)

(continued)

Table 3.1 (continued)

Plant part	Compounds	References
Flower	Kaempferol, $C_{15}H_{10}O_6$	Gopal and Gupta (1972)
	Quercetin	
	Free β-sitosterol	
	β-D-glucoside of β-sitosterol	
	Hentriacontane	
	Hentriacontanol	
	Pelargonidin-5-beta-glucopyranoside	Niranjan and Gupta (1973)
	Cyanidin 7-methyl ether-3-β-glucopyranoside	
	24β-ethylcholest-5-en-3 β-O-α-L- arabinopyranosyl $(1 \rightarrow 6)$-β-D-glucopyranoside [9]	Rizvi and Saxena (1974)
	3,5-dihydroxy-4-methoxyflavone-7-O-α-L-rhamnopyranosyl$(1 \rightarrow 6)$-β-D-glucopyranoside [10]	
	4,5,7-trihydroxyflavone-3-O-β-D-glucopyranosyl $(1 \rightarrow 4)$-α-L-rhamnopyranoside [11]	
	3,5,7-trihydroxy-4'-methoxyflavone, 3,5,7,4'-tetrahydroxy-3'-methoxyflavone	
	Kaempferol, Kaempferide-7β-rutinoside	
	Quercetin, Quercetin-3β-rutinoside, Isorhamnetin-3-β-rutinoside	
	Ceryl alcohol $C_{26}H_{54}O$, β-amyrin, $C_{30}H_{50}O$	
	Taraxerol, Taraxryl acetate	
	Lupeol [3], Lupeol acetate	
	Fatty acids: Myristic, Palimitic, Stearic, Arachidic, Behenic, Oleic, Linoleic, Linolenic, Arachidonic acid	

(continued)

Table 3.1 (continued)

Plant part	Compounds	References
	Sugars: Sucrose, L-rhamnose, D-xylose, L-arabinose, D-fructose, D-glucose, D-galactose, 3-O-β-L-arabinopyranosyl-L-arabionse, 4-O-β-D-galactopyranosyl-D-galactose, 4-O-β-D-glucopyranosyl-L-rhamnose, 4-O-β-D-glucopyranosyl-D-glucose, 3-O-β-D-galactopyranosyl-L-arabinose, 6-O-α-L-rhamnopyranosyl-D-glucose, 6-O-α-L-rhamnopyranosyl-D-galactose, 6-O-β-D-galactopyranosyl-D-glucose, O-β-D-galactopyranosyl-(1 → [4-O-β-D-galactopyranosyl-1]₂ → O-D-galactose	
	Hexadecanoic acid, Tetradecanoic acid	Wang et al. (2003)
	β-cedrol, α-cedrol	
	3-methyl-2(3H)-benzofuranone	
	Bombasin: (= 1-[(2R,3S)-2,3-dihydro-2-(4-hydroxy-3-methoxy-phenyl)-3-(hydroxymethyl)-7-methoxybenzofuran-5-yl]ethanone), $C_{19}H_{20}O_6$	Wu et al. (2008)
	Bombasin 4-O- β-glucoside:(= 1-[2R,3S)-2,3-dihydro-2-(4-β-glucopyranosyloxy-3-methoxyphenyl)-3-(hydroxymethyl)-7-methoxybenzofuran-5-yl]ethanone), $C_{25}H_{30}O_{11}$	
	Bombalin: (= (2S)-2-hydroxy-3-[(2S,3S,4R)-4-hydroxy-3-methoxy-5-oxotetrahydrofuran-2-yl]ethyl(2E)-3-(4-hydroxyphenyl)prop-2-enoate, $C_{16}H_{18}O_8$	
	Dihydrodehydrodiconiferyl alcohol 4-O-β-D-glucopyranoside [12]	
	Trans-3-(p-coumaroyl) quinic acid [13]	
	Neochlorogenic acid [14]	
	Vicenin 2: Apigenin-6, 8-di-C-β-D-glucopyranoside, $C_{27}H_{30}O_{15}$ [15]	El-Hagarassi et al. (2011)

(continued)

Table 3.1 (continued)

Plant part	Compounds	References
	Linarin: Apigenin-4′-methylether-7-O-rutinoside, $C_{28}H_{32}O_{14}$ [16]	
	Saponarin: Apigenin-6-C-β-D-glucosyl-7-O-β-D-glucopyranoside, $C_{27}H_{30}O_{15}$ [17]	
	Cosmetin: Apigenin-7-O-β-D-glucopyranoside, $C_{21}H_{20}O_{10}$ [18]	
	Isovitexin:Apigenin-6-C-β-D-glucopyranoside, $C_{21}H_{20}O_{10}$ [19]	
	Xanthomicrol: 4′,5-dihydroxy-6,7,8-trimethoxyflavone, $C_{18}H_{16}O_7$ [20]	
	Apigenin, $C_{15}H_{10}O_5$ [21]	
	Cholesterol	
	Campesterol	
	Stigmasterol	
	α-amyrin	
	Behenic, Capric, Arachidic, Myristic, Stearic, Linoleic, Pentadecandienoic acid	
	Isovanilic acid:4-methoxy-3-hydroxybenzoic acid	Said et al. (2011)
	Mangiferin [8]	
	Protocatechuic acid:3,4 dihydroxy-benzoic acid	
	Rutin	
	Quercetin 3-O-β-D-glactouronopyranoside	
	Quercetin 3-O-β-D-glucopyranoside	
	Apigenin-7-O-β-D-glucopyranoside	
Seed	Crude Protein and Pentosan	Kapoor et al. (1975)
	Gallic acid	Dhar and Munjal (1976)
	Tannic acid	
	Ethyl gallate	
		(continued)

Table 3.1 (continued)

Plant part	Compounds	References
	1-galloyl-β-glucose	Rastogi and Mehrotra (1995)
	n-hexacosanol	
	Palmitic acid, Octadecyl palmitate	
	A mixture of α,β and γ-tocopherols	
	Carotenoids	
	Fatty acids: Palmitic, Linoleic, Behenic, Myristic, Arachidic acid	
	β-sitosterol	
	α-Tocopherol	
	Glucose, Xylose, Rhamnose	
	n-hexacosanol	
	Amino acids: Alanine, Valine, Isoleucine, Leucine, Arginine, Glycine, Aspartic acid	
Gum	Riboflavin	Broker and Bhat (1953)
	Thiamine	
	2,3,4,6-tetra-, 2,6-di- and 2,4-di-O-methyl-D-galactose	Bose and Dutta (1963a,b); Bose and Dutta (1965)
	2,3,5-tri and 2,5-di-O-methyl-L-arabinose	
	D-galactose	
	L-arabinose	
	Rhamnose	
	D-galacturonic acid	

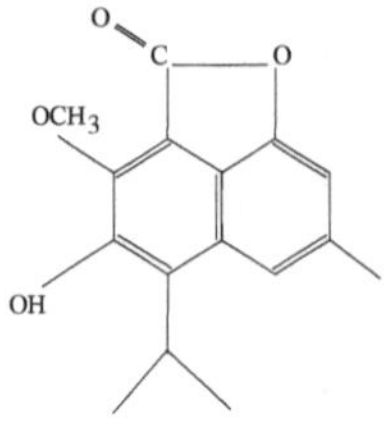

Fig. 3.1 Chemical structures of some compounds isolated from *Bombax ceiba*

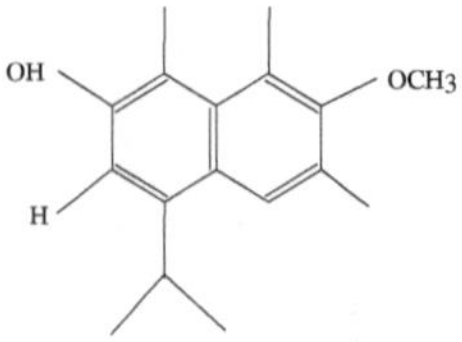

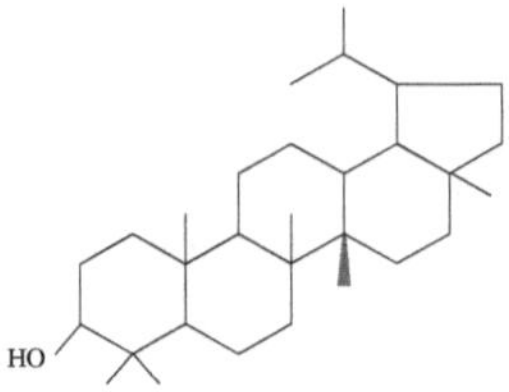

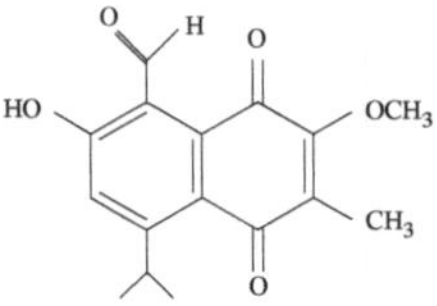

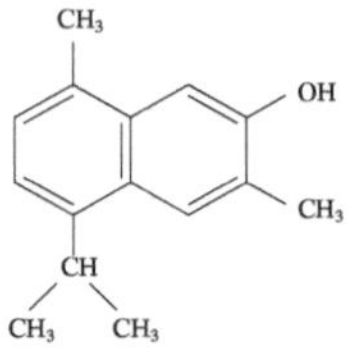

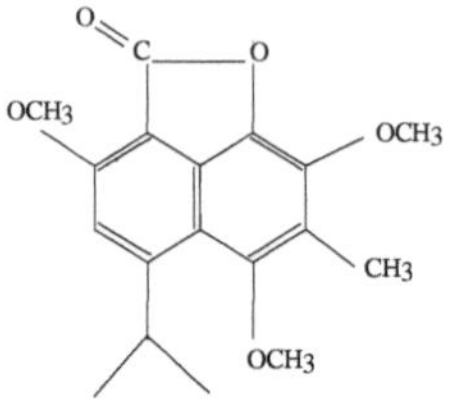

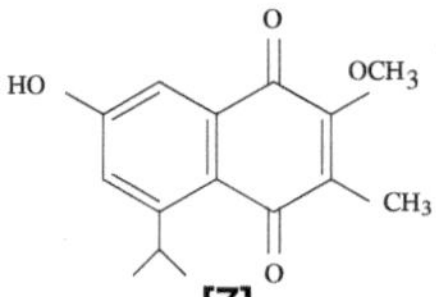

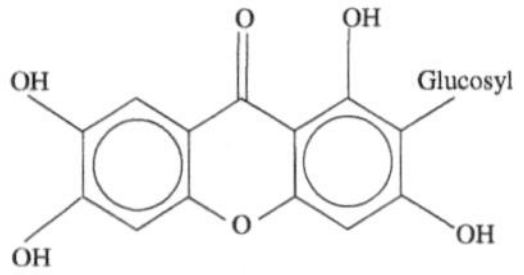

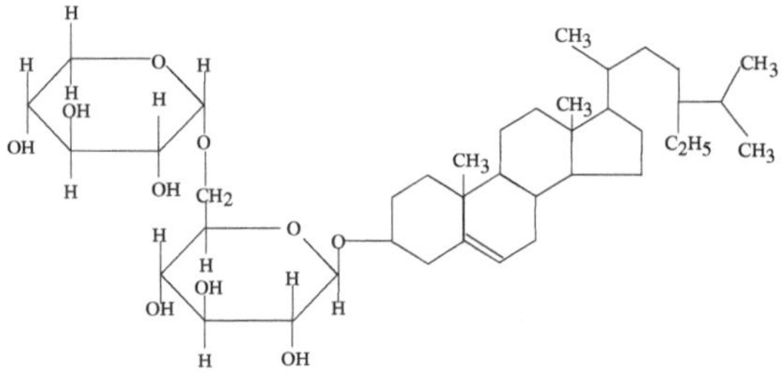

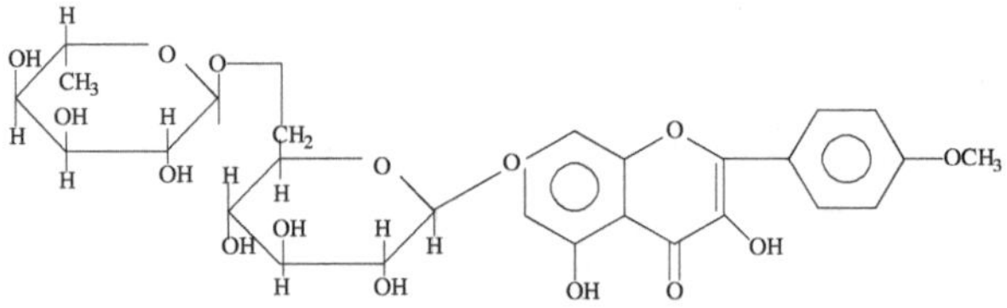

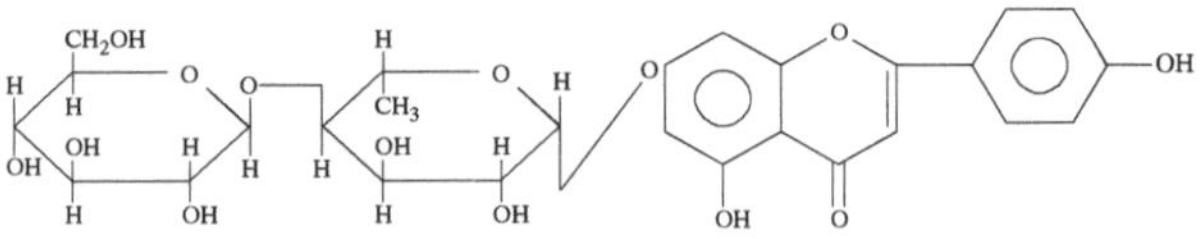

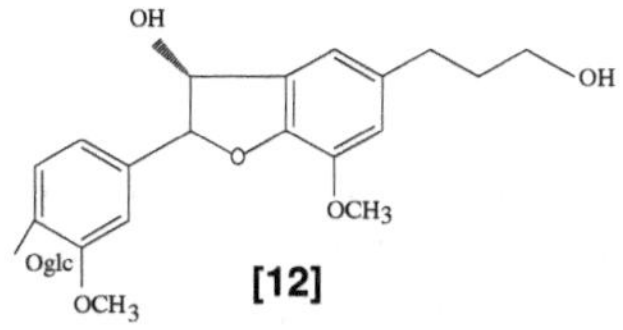

Fig. 3.1 (continued)

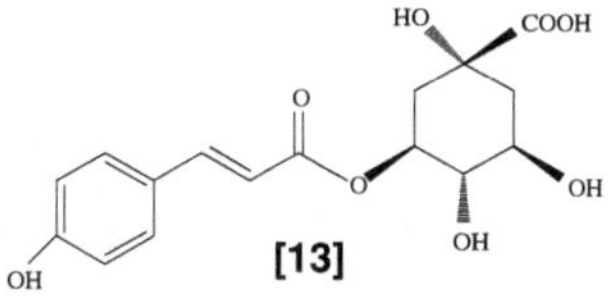

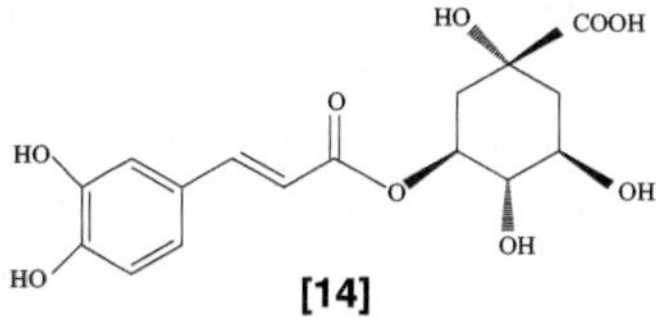

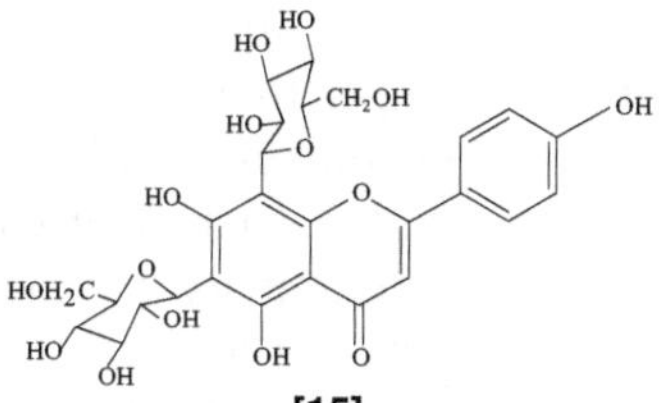

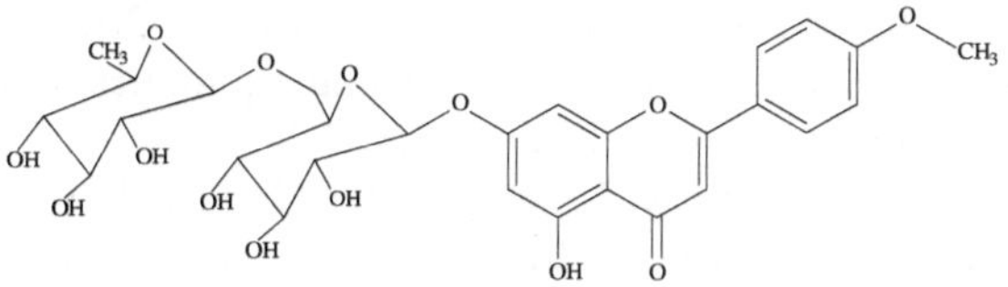

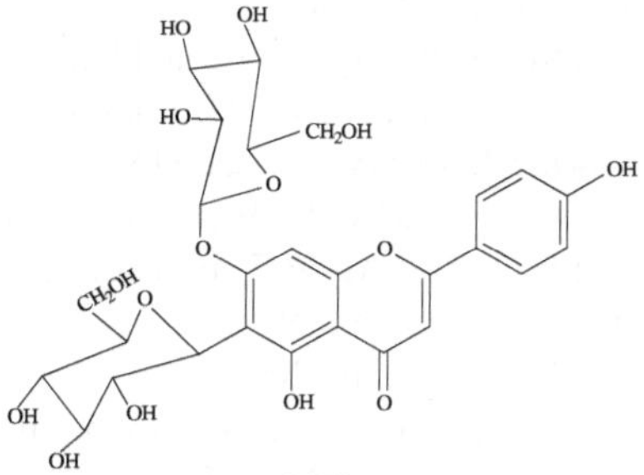

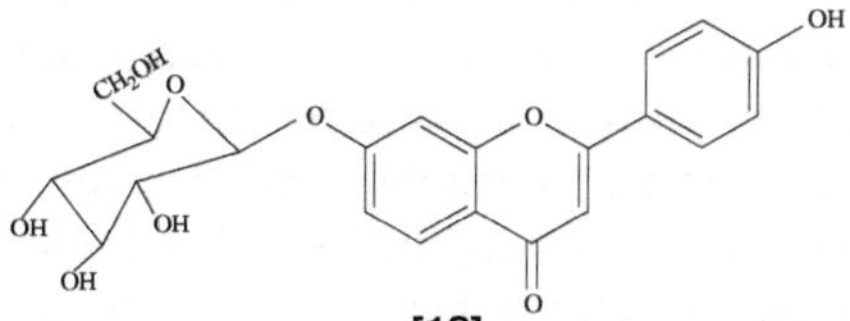

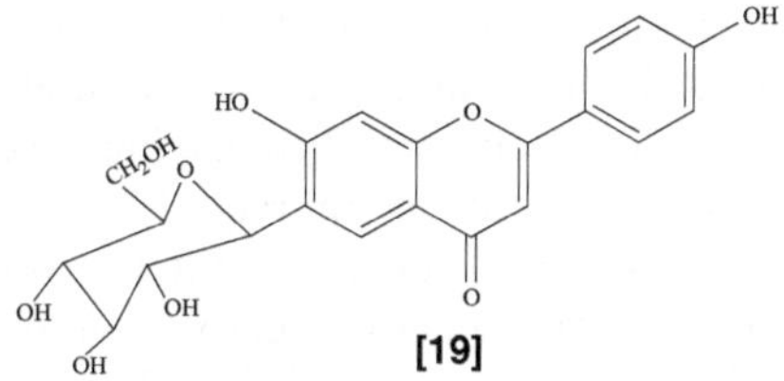

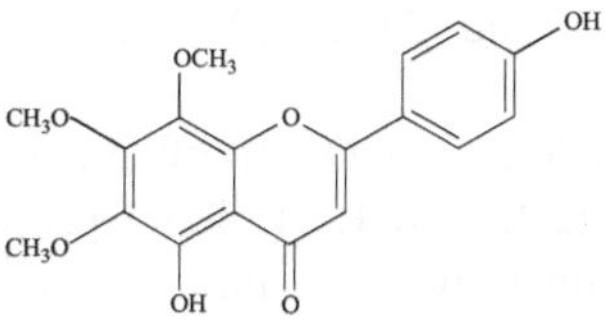

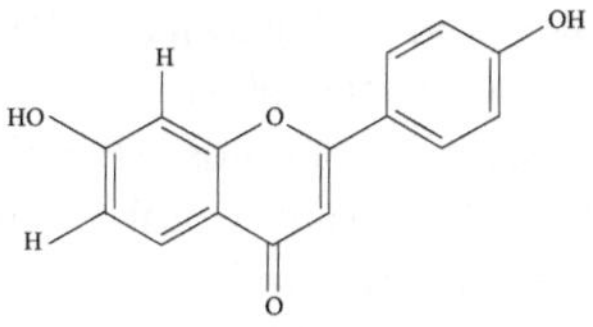

3.2 Root

Chemical analysis of barkless, white pulpy roots of two-year old plants has revealed to contain 71.2% starch, 8.2% sugars (arabinose and galactose), 1.2% proteins, 0.4% tannins, 0.1% nontannins, 0.9% fat, 2.1% mineral matter, 2.0% cellulose, 7.5% moisture, 6.0% pectic substances, 0.3% phosphatide-cephalin and mucilage. Younger roots contain more sugars, starch and pectic substances than older roots, but contain less oil, coloring matter and cellulose (Ghose 1935). Roots also possess higher content of calcium (93 mg/100 g) as shown by Ghate et al. (1988). In a preliminary phytochemical study, roots have shown presence of flavonoids, tannins, saponins, steroids, cardiac-glycosides, phenols besides carbohydrates and amino acids. Total phenolic content of dried root powder was 4.86% and total tannin content was 1.72% (Jain et al. 2011). Detailed chemical investigations led to isolation of various naphthoquinones, sesquiterpenoids and phenolic compounds in the roots (Tables 3.1, 3.2).

Isohemigossylic acid lactone 2-methyl ether [2] has shown antifungal activity by inhibiting conidia growth of *Verticillium dahliae* fungal strain V76 in a turbimetric bioassay. It had an inhibitory effect on conidia over the range of 0.5–30.0 µg/ml and ED_{50} value obtained was 7.8 ± 2.3 µg/ml (Puckhaber and Stipanovic 2001). 2-O-methylisohemigossylic acid lactone, a sesquiterpene, displayed strong growth inhibitory effect against human promyelotic leukemia HL-60 cells (Hibasami et al. 2004). 7-hydroxy cadalene [5] has shown strong antibacterial potential against *Bacillus subtilis* (MIC-0.59 µg/ml). However, it was found to be less active ($IC_{50} > 5$ µg/ml) against MCF-7 (breast adenocarcinoma), KB (human oral cancer), HeLa (human cervical cancer) and HT-29 (colon cancer) cell lines in sulforhodamine B assay (Boonsri et al. 2008).

Daucosterol, a β-sitosterol glycoside, has shown immunomodulatory activity against disseminated candidiasis caused by *Candida albicans* (Lee et al. 2007). Hesperidin is an antioxidant and anti-allergic compound (Balakrishnan and Menon 2007; Jeong et al. 2011). It also inhibits bone loss in ovariectomized mice (Chiba et al. 2003) and significantly increases HDL and lowers cholesterol, LDL-C and triglycerides levels in rats (Monforte et al. 1995). Oleanolic acid is another interesting molecule present in the root which possesses many biological activities, for example hepatoprotective, antihyperlipidemic, anti-inflammatory, antitumor (Liu 1995), cardioprotective (Martínez-González et al. 2008), antioxidant, antiglycative (Balanehru and Nagarajan 1991; Yin and Chan 2007), antibacterial and antiparasitic (Szakiel et al. 2008).

Some novel sesquiterpenoids (Bombamalones A-D, Bombamaloside, Isohemigossypol-1-methyl ester, Bombaxquinone B and Lacinilene) present in the root were screened for cytotoxicity assay against HGC-27 human gastrointestinal cancer cell line using a 3-(4,5-dimethylthiazol-2-yl)-2,5-dipehnyl-2H-tetrazolium bromide (MTT) with Paclitaxel as a positive control. However, all were found to be inactive. Bombamalone B [10-hydroxy-3,5,5,10-tetramethylbenzofuro(4,3-*fg*)chromene-1,2(5H,10H)-dione] was also checked for its activity against A549

Table 3.2 Distribution of compounds isolated from *B. ceiba* in some major chemical classes

Compounds	Isolated from plant part	References
Flavonoids		
3,5-dihydroxy-4-methoxyflavone-7-O-α-L-rhamnopyranosyl(1 → 6)-β-D-glucopyranoside.	Flower	Niranjan and Gupta (1973); Rizvi and Saxena (1974); El-Hagarassi et al. (2011); Said et al. (2011)
4,5,7-trihydroxyflavone-3-O-β-D-glucopyranosyl (1 → 4)-α-L-rhamnopyranoside		
3,5,7-trihydroxy-4′-methoxyflavone		
3,5,7,4′-tetrahydroxy-3′-methoxyflavone		
Kaempferol		
Kaempferide-7β-rutinoside		
Quercetin		
Quercetin-3β-rutinoside		
Isorhamnetin-3-β-rutinoside		
Apigenin		
Vicenin 2		
Linarin		
Saponarin		
Cosmetin		
Isovitexin		
Rutin		
Isovanillic acid		
Xanthomicrol		
Quercetin 3-O-β-D-glactouronopyranoside		
Quercetin 3-O-β-D-glucopyranoside		
Apigenin-7-O-β-D-glucopyranoside		
Pelargonidin-5-β-glucopyranoside		
Cyanidin 7-methylether-3-β-glucopyranoside		

(continued)

Table 3.2 (continued)

Compounds	Isolated from plant part	References
Mangiferin	Leaves	Shahat et al. (2003)
	Flower	Said et al. (2011)
Shamimicin:1,1-bis-2- (3,4-dihydroxyphenyl)-3,4-dihydro-3,7-dihydroxy-5-O-xylopyranosyloxy-2H-1-benzopyran	Stem bark	Saleem et al. (2003)
Hesperidin	Root	Qi et al. (1996)
7-hydroxy cadalene	Root bark	Sankaram et al. (1981)
	Heart-wood	Sreeramulu et al. (2001)
Phenolic compounds		
1,6-dihydroxy-3-methyl-5-(1-methylethyl)-7-methoxy-8-carboxylic acid (8 → 1 lactone)	Root	Sood et al. (1982); Sankaram et al. (1981); Puckhaber and Stipanovic (2001); Reddy et al. (2003)
Isohemigossylic acid lactone 2-methyl ether.		
Isohemigossypol-1-methyl ether		
5-isopropyl-3-methyl-2,4,7 trimethoxy-8,1-naphthalene carbolactone		
8-formyl-7-hydroxy-5-isopropyl-2-methoxy-3-methyl-1, 4-napthaquinone		
7-hydroxy-5-isopropyl-2-methoxy-3-methyl-1,4-naphthoquinone	Heart wood	Sreeramulu et al. (2001)
8-formyl-7-dydroxy-5-isopropyl-2-methoxy-3-methyl-1,4-naphthoquinone		
Trans-3-(p-coumaroyl) quinic acid	Flower	Wu et al. (2008)
Neochlorogenic acid		Said et al. (2011)
Protocatechuic acid		
Gallic acid	Seed	Dhar and Munjal (1976)
Tannic acid		
Steroids and Terpenoids		

(continued)

Table 3.2 (continued)

Compounds	Isolated from plant part	References
β-sitosterol	Root	Chauhan et al. (1980)
	Root bark	Sheshadri et al. (1971)
	Stem bark	Mukherjee and Roy (1971)
	Flower	Gopal and Gupta (1972)
	Seed	Rastogi and Mehrotra (1995)
Lupeol	Root bark	Sheshadri et al. (1971)
	Stem bark	Mukherjee and Roy (1971)
	Flower	Rizvi and Saxena (1974)
Daucosterol	Root	Qi et al. (1996); Zhang et al. (2007)
Oleanolic acid		
Bombamalone A: 3,7-dihydroxy-5-isopropyl-8-methoxy-7-methyl-7H-naphthol [1,8-bc] furan-2,6-dione		
Bombamalone B: 10-hydroxy-3,5,5,10-tetramethylbenzofuro [4,3-fg] chromene-1,2(5H, 10H)-dione		
Bombamalone C: 7-hydroxy-4-isopropyl-8-methoxy-6-methyl naphthalenene-1,2-dione		
Bombamalone D: 5,8-dihydro-2-hydroxy-4-isopropyl-7-methoxy-6-methyl-5,8-dioxonaphthalene-1- carboxylic acid		
Bombamaloside:4-O-β-glucopyranosyl-6,7-dihydro-2,2,8-trimethyl-6,7-dioxo-2H-naphthol [1,8-bc] furan-5-carboxylic acid		
Lacinilene		
Bombaxquinone B		

(continued)

Table 3.2 (continued)

Compounds	Isolated from plant part	References
24β-ethylcholest-5-en-3 β-O-α-L-arabinopyranosyl (1 → 6)-β-D-glucopyranoside	Flower	Rizvi and Saxena (1974)
Cholesterol		El-Hagrassi et al. (2011)
Campesterol		
Stigmasterol		
α-amyrin		
β-amyrin		
Taraxerol		
Taraxryl acetate		
Lupeol acetate		
α-, β- and γ-tocopherols	Seed	Dhar and Munjal (1976)
Carotenoids		
Hydrocarbons and fatty acids		
Myristic acid	Flower	Rizvi and Saxena (1974)
	Seed	Rastogi and Mehrotra (1995)
Palimitic acid	Flower	Rizvi and Saxena (1974)
	Seed	Rastogi and Mehrotra (1995)
Stearic acid	Flower	Rizvi and Saxena (1974)
Arachidic acid	Flower	Rizvi and Saxena (1974)
	Seed	Rastogi and Mehrotra (1995)
Behenic acid	Flower	El-Hagarassi et al. (2011)
	Seed	Rastogi and Mehrotra (1995)
Capric acid	Flower	El-Hagarassi et al. (2011)
Linoleic acid	Flower	El-Hagarassi et al. (2011)
	Seed	Rastogi and Mehrotra (1995)
Pentadecandienoic acid	Flower	El-Hagrassi et al. (2011)

(continued)

Table 3.2 (continued)

Compounds	Isolated from plant part	References
Hentriacontanol	Flower	Gopal and Gupta (1972)
Hentriacontane		
n-hexacosanol	Seed	Dhar and Munjal (1976)
n-triacontanol	Root	Chauhan et al. (1980)
Carbohydrates		
Sucrose	Flower	Rizvi and Saxena (1974)
L-rhamnose	Flower	Rizvi and Saxena (1974)
D-xylose	Seed	Rastogi and Mehrotra (1995)
L-arabinose	Flower	Rizvi and Saxena (1974)
D-fructose	Seed	Rastogi and Mehrotra (1995)
D-glucose	Flower	Rizvi and Saxena (1974)
D-galactose	Gum	Bose and Dutta (1963a,b)
3-O-β-L-arabinopyranosyl-L-arabionse	Flower	Rizvi and Saxena (1974)
4-O-β-D-galactopyranosyl-D-galactose	Flower	Rizvi and Saxena (1974)
4-O-β-D-glucopyranosyl-L-rhamnose	Seed	Rastogi and Mehrotra (1995)
4-O-β-D-glucopyranosyl-D-glucose	Flower	Rizvi and Saxena (1974)
3-O-β-D-galactopyranosyl-L-arabinose	Gum	Bose and Dutta (1963a, b)
6-O-α-L-rhamnopyranosyl-D-glucose	Flower	Rizvi and Saxena (1974)
6-O-α-L-rhamnopyranosyl-D-galactose		
6-O-β-D-galactopyranosyl-D-glucose		
O-β-D-galactopyranosyl-(1 → [4-O-β-D-galactopyranosyl-1]$_2$ → O-D-galactose		

(continued)

Table 3.2 (continued)

Compounds	Isolated from plant part	References
D-galacturonic acid	Gum	Bose and Dutta (1963a, b); Bose and Dutta (1965)
2,3,4,6-tetra-O-methyl-D-galactose		
2,6-di-O-methyl-D-galactose		
2,4-di-O-methyl-D-galactose		
2,3,5-tri -O-methyl- L-arabinose		
2,5-di-O-methyl- L-arabinose		
9′-normeolignan		
Bombasin: (= 1-[(2R,3S)-2,3-dihydro-2-(4-hydroxy-3-methoxy-phenyl)-3-(hydroxymethyl)-7-methoxybenzofuran-5-yl]ethanone	Flower	Wu et al. (2008)
Bombasin 4-O-glucoside:(= 1-[2R,3S)-2,3-dihydro-2-(4-β-glucopyranosyloxy-3-methoxyphenyl)-3-(hydroxymethyl)-7-methoxybenzofuran-5-yl]ethanone)		
Bombalin: (= (2S)-2-hydroxy-3-[(2S,3S,4R)-4-hydroxy-3-methoxy-5-oxotetrahydrofuran-2-yl]ethyl(2E)-3-(4-hydroxyphenyl)prop-2-enoate		
Dihydrodehydrodiconiferyl alcohol 4-O-β-D-glucopyranoside		

lung carcinoma, MCF-7 breast carcinoma and HeLa cervical human cancer cell lines and was found to be inactive (Zhang et al. 2007).

β-sitosterol isolated from both root and root bark possess antihyperglycemic (Ivorra et al. 1988), analgesic, antihelminthic, antimutagenic (Villaseñor et al. 2002), angiogenic (Choi et al. 2002) and chemopreventive (Baskar et al. 2010) activities. It significantly inhibits vascular adhesion molecule 1 and intracellular adhesion molecule 1 expression in TNF alpha-stimulated HAEC as well as the binding of U937 cells to TNF-alpha-stimulated HAEC and attenuates the phosphorylation of nuclear factor-kB p65 (Loizou et al. 2010) and decreases cholesterol synthesis at the level of HMG-CoA reductase gene expression (Field et al. 1997). Lupeol [3], a pentacylic triterpene has demonstrated anti-inflammatory, antiarthritic, antiangiogenic, hepatoprotective, hypotensive, cardioprotective and antileukemia activities (You et al. 2003; Wal et al. 2011).

3.3 Stem Bark

Stem-bark possesses 9.92% total water extractive, 3.01% tannins and 6.91% nontannin (Chadha 1972). Shamimicin, a flavanoid isolated from stembark was screened for its hypotensive activity in animal model; however, it did not reduce mean arterial blood pressure significantly at the dose of 15 mg/kg (Saleem et al. 2003).

3.4 Leaf

Leaves are reported to contain condensed type of tannins (Chadha 1972). However, studies on chemistry of its leaves are limited. Mangiferin [8], a flavanoid has been isolated from leaves which was earlier misinterpreted as a novel compound Shamimin (Faizi and Ali 1999) and later correctly interpreted by Shahat et al. (2003) to be Mangiferin. Mangiferin has demonstrated significant hypoglycemic, hypotensive, hypolipidemic, anti-HIV-1, analgesic, antioxidant and antimicrobial activities (Saleem et al. 1999; Dar et al. 2005; Muruganandan et al. 2005; Stoilova et al. 2008; Wang et al. 2011). Recently, in a preliminary phytochemical screening, Hossain et al. (2011) have demonstrated presence of steroids, carbohydrates, tannins, triterpenoids, deoxysugar, flavonoids and coumarin glycosides in methanolic extract of leaves.

Table 3.3 Chemical composition of flower buds and calyces on fresh water basis (Bhatnagar 1965)

Composition	Flower buds	Calyces
Moisture (%)	85.66	85.14
Carbohydrates (%)	11.95	13.87
Crude protein (%)	1.38	1.56
Mineral matter (%)	1.09	1.00
Ether extract (%)	0.44	0.51
Calcium (mg/100 g)	92.25	95.0
Phosphorus (mg/100 g)	49	41
Magnesium (mg/100 g)	54.24	64

3.5 Flower

Bhatnagar (1965) studied the chemical composition of flowers and calyx of *B. ceiba* (Table 3.3). It has demonstrated that protein and phosphorus content, and ether extract of raw calyces compare favorably with those of the common vegetables such as carrots, radishes, turnips, pumpkin and cabbages.

Detailed chemical studies of flowers have shown that they contain various types of flavonoids (Table 3.2) of which Apigenin, Quercetin and Kaempferol are well known for their beneficial biological activities.

Apigenin **[21]** possesses antineoplastic, antispasmolytic, antihypertensive activities (Nagarathnam and Cushman 1991; Rajnarayana et al. 2001). It is an effective proteasome inhibitor in cultured breast cancer cells and in breast cancer xenografts. Furthermore, it induces apoptotic cell death in human breast cancer cells and exhibits anticancer activities in tumors (Chen et al. 2007). Apigenin also inhibits glucosyltransferases and phosphodiesterase activity (Rajnarayana et al. 2001; De Sanchez et al. 1996; Koo et al. 2002). Kaempferol has shown in vitro fibrinolytic activity (Rajput et al. 2011) and exerts a potent inhibitory effect on in vitro bone resorption (Lorget et al. 2003) along with estrogenic activity (Zoechling et al. 2009). It is a potent chemopreventive agent against skin cancer through its inhibitory interaction with Src (Lee et al. 2010). Kaempferol-3-O-rutinoside has shown strong in vitro alpha-glucosidase inhibitory activity as compared to reference antidiabetic drug, acarbose (Habtemariam 2011).

Quercetin possesses immunomodulatory, hypoglycemic, antioxidant, antiulcer, anti-inflammatory, antibacterial, antiviral, positive ionotropic (Okoko and Or-uambo 2009; Itoigawa et al. 1999; Igarashi and Ohmuna 1995; McAnlis et al. 1997; Rajnarayana et al. 2001; Izzo et al. 1994; Robak and Glyglewski 1988; Fuhrman et al. 1995) and inhibitory activity on in vitro bone resorption, aldose reductase, tyrosine kinase, Na^+/K^+ATPase, H^+ATPase, aryl hydroxylase and epoxide hydroxylase (Wattel et al. 2003; Lorget et al. 2003; Rajnarayana et al. 2001).

Cosmetin **[18]** possesses anti-HIV, antioxidant, reflux oesophagitis and gastritis inhibiting activities (Tang et al. 1994; Min et al. 2005). Vitexin is reported to possess antiplasmodial, antitrypanosomal and antileishmanicidal activities (Lagnika et al. 2009). Vicenin-2 [15] has shown antioxidant, anticancer, antihepatotoxic, anti-inflammatory and antitrypanosidal activities (Velozo et al. 2009; Hoffmann-Bohm et al. 1992; Grael et al. 2005; Marrassini et al. 2011; Nagaprashantha et al. 2011). Saponarin [17] possesses antioxidant, hypoglycemic and hepatoprotective properties (Sengupta et al. 2009; Vitcheva et al. 2011). Linarin [16] possesses sedative, sleep-enhancing, acetylcholinesterase and NO inhibitory activities (Fernández et al. 2004; Fan et al. 2008; Shinha et al. 2002). Xanthomicrol [20] possesses spasmolytic and cytotoxic activities (Jahaniani et al. 2005; Meckes et al. 2002). Cyanidin-3-glucoside (C3G) enhanced neurite outgrowth by promoting p-Glycogen Synthase Kinase 3b(Ser9) and reversed ethanol-mediated activation of GSK3b and inhibition of neurite outgrowth as well as the expression of NF proteins. C3G also blocked ethanol-induced intracellular accumulation of reactive oxygen species (Chen et al. 2009).

Taraxerol has reported to have analgesic (Biswas et al. 2009), antibacterial and cytotoxic (Online document 1) activities. Alpha- and Beta-amyrin possess analgesic, anti-inflammatory, antinociceptive, sedative, anxiolytic and platelet aggregation inhibition activities (Aragão 2004; Aragao et al. 2008). Alpha-amyrin also possesses antihyperglycemic (Singh et al. 2009) and beta-amyrin possesses antidepressant activity (Subarnas et al. 1993).

Hentriacontane has been reported to ameliorate the expression of inflammatory mediators (TNF-α, IL-6, PGE(2), COX-2 and iNOS) and the activation of NF-$\kappa\beta$ and caspase-1 in LPS-stimulated peritoneal macrophages (Kim et al. 2011b). It has been reported to reduce the virulence of experimentally induced *Plasmodium vinckei* and can be developed as a new antimalarial compound (Deharo et al. 1992).

Gallic acid is reported to possess antianxiety (Dhingra et al. 2011), anti-inflammatory (Kroes et al. 1992), analgesic (Krogh et al. 2000) and antiangiogenic (Liu et al. 2006) activities. It is a histone acetyltransferase inhibitor, and suppresses β-amyloid neurotoxicity by inhibiting microglial-mediated neuroinflammation (Kim et al. 2011a).

Bombasin, Bombasin 4-O-glucoside (9-norneolignans) and a novel D-gluconolactone derivative Bombalin were found to be inactive (IC$_{50}$ > 50 M) against HGC-27 gastrointestinal and HeLa cervical human cancer cell lines (Wu et al. 2008).

3.6 Seed

Seeds of *B. ceiba* (Indian Kapok) yield pale-yellow colored oil (22.3% yield) which deposits stearin on standing. The oil from its seeds possesses higher amount of saturated fatty acids than those of true Kapok (*Ceiba pentandra*) oil. It contains 1.2% myristic, 23.6% palmitic, 64.9% oleic, 7.5% linoleic and 2.8% arachidic acids. Oil has shown an iodine value of 78.0 which is quite lower than of true

Kapok oil (117.9). Specific gravity of oil at 15° C is 0.9208 with an acid value of 9.3, Saponification value of 193.3 and Reichert-Meissl value of 0.5. Chemical analysis of its seed meal has shown that it has 11.40% moisture, 36.50% protein, 0.80% fat, 24.70% carbohydrates, 19.90% crude fiber, 24.70% mineral matter and a nutritive ratio of 1:0.7 (Chadha 1972).

n-hexacosanol, a cyclohexenonic long chain saturated fatty alcohol, exerts neurotrophic properties on central neurons and stimulates phagocytosis in macrophages (Borg 1991). It has also shown an inhibitory effect on insulin secretion stimulated by glucose in vivo and in vitro in the rat (Damgé et al. 1995). In another animal study, n-hexacosanol (8 mg/kg, i.p.) improved diabetes-induced nitric oxide synthase alterations in the kidney, resulting in the amelioration of diabetic nephropathy (Okada et al. 2008). It has therapeutic effects on hypercontractility in the diabetic ileum by ameliorating over expression of muscarinic M_2 and M_3 receptors mRNAs as shown in an animal study (Shinbori et al. 2006).

3.7 Gum

Stem bark exudes a dark-brown colored gum known as *Mocharus/Semal-gum* from the natural wounds, caused by decay or by insects, or as a result of some functional disease. It is almost insoluble in water but absorbs it and swells like true gum tragacanth. Purified gum possesses 8.9% mineral matter and a considerable amount of catechol tannin. Complete hydrolysis of gum has revealed that it contains a mixture of various sugars and gallic and tannic acids (Bose and Dutta 1963a, b).

References

Aragão GF (2004) Antiinflammatory, antiaggregant activity and central effects of alpha and beta amyrin from *Protium heptaphyllum* Aubl March. http://www.openthesis.org/documents/Antiinflammatory-antiaggregant-activity-central-effects-332976.html. Accessed 2 Nov 2011

Aragao GF, Pinheiro MCC, Bandeira PN, Lemos TLG, Viana GSB (2008) Analgesic and Anti-Inflammatory Activities of the Isomeric Mixture of Alpha- and Beta-Amyrin from *Protium heptaphyllum*(Aubl.) March. J Herbal Pharmacother 7(2):31–47

Balakrishnan A, Menon VP (2007) Antioxidant properties of hesperidin in nicotine-induced lung toxicity. Fundam Clin Pharmacol 21(5):535–546

Balanehru S, Nagarajan B (1991) Protective effect of oleanolic acid and ursolic acid against lipid peroxidation. Biochem Int 24(5):981–990

Baskar AA, Ignacimuthu S, Paulraj GM, Numair KL (2010) Chemopreventive potential of β-Sitosterol in experimental colon cancer model—an *In vitro* and *In vivo* study. BMC Complement Alternat Med 10:24. doi:10.1186/1472-6882-10-24

Bhatnagar MS (1965) Chemical composition of calyx of *Salmalia malabarica*. Sci Cult 31:189

Biswas M, Biswas K, Ghosh AK, Haldar PK (2009) A pentacyclic triterpenoid possessing analgesic activity from the fruits of *Dregea volubilis*. Phcog Mag 5:90–92

Boonsri S, Karalai C, Ponglimanont C, Chantrapromma S, Kanjana-opas A (2008) Cytotoxic and antibacterial sesquiterpenes from *Thespesia populnea*. J Nat Prod 71:1173–1177

Borg J (1991) The neurotrophic factor, n-hexacosanol, reduces the neuronal damage induced by the neurotoxin, kainic acid. J Neurosci Res 29(1):62–67

Bose S, Dutta AS (1963a) Structure of *Salmalia malabarica* gum. I. Nature of sugars present and the structure of aldobiuronic acid. J Indian Chem Soc 40:257–262

Bose S, Dutta AS (1963b) The structure of *Salmalia malabarica* gum. II. Structure of the degraded gum. J Indian Chem Soc 40:557–561

Bose S, Dutta AS (1965) The structure of *Bombax malabarica* gum. III. A tentative structure from methylation studies. J Indian Chem Soc 42:367–372

Broker R, Bhat JV (1953) Riboflavin and thiamine contents of gums. Curr Sci 22:343

Chadha YR (1972) The Wealth of India, raw material, vol IX. Publications and Information Directorate, New Delhi

Chauhan JS, Sultan M, Srivastav SK (1980) Constituents from *Salmalia malabaricum*. Can J Chem 58:328–330

Chen D, Landis-Piwowar KR, Chen MS, Dou QP (2007) Inhibition of proteasome activity by the dietary flavonoid apigenin is associated with growth inhibition in cultured breast cancer cells and xenografts. Breast Cancer Res 9(6):R80

Chen G, Bower KA, Xu M, Ding M, Shi X, Ke Z-J, Luo J (2009) Cyanidin-3-glucoside reverses ethanol-induced inhibition of neurite outgrowth: role of glycogen synthase kinase 3 beta. Neurotoxic Res 15:321–331

Chiba H, Uehara M, Wu J, Wang X, Masuyama R, Suzuki K, Kanazawa K, Ishimi Y (2003) Hesperidin, a *Citrus* flavonoid, inhibits bone loss and decreases serum and hepatic lipids in ovariectomized mice. J Nutr 133(6):1892–1897

Choi S, Kim KW, Choi JS, Han ST, Park YI, Lee SK, Kim JS, Chung MH (2002) Angiogenic activity of beta-sitosterol in the ischaemia/reperfusion-damaged brain of Mongolian gerbil. Planta Med 68(4):330–335

Damgé C, Hillaire-Buys D, Koenig M, Gross R, Hoeltzel A, Chapal J, Balboni G, Borg J, Ribes G (1995) Effect of n-hexacosanol on insulin secretion in the rat. Eur J Pharmacol 274(1–3):133–139

Dar A, Faizi S, Naqvi S, Roome T, Zikr-ur-Rehman S, Ali M, Firdous S, Moin ST (2005) Analgesic and antioxidant activity of mangiferin and its derivatives: the structure activity relationship. Biol Pharm Bull 28(4):596–600

Deharo E, Sauvain M, Moretti C, Richard B, Ruiz E, Massiot G (1992) Antimalarial effect of n-hentriacontanol isolated from *Cuatresia* sp (Solanaceae). Ann Parasitol Hum Comp 67(4):126–127

Dhar DN, Munjal RC (1976) Chemical examination of the seeds of *Bombax malabaricum*. Planta Med 29:148–150

Dhingra D, Chhillar R, Gupta A (2011) Antianxiety-like activity of gallic acid in instressed and stressed Mice: possible involvement of nitriergic system. Neurochem Res doi:10.1007/S11064-011-0635-7(EPub)

El-Hagrassi AM, Ali MM, Osman AF, Shaaban M (2011) Phytochemical investigation and biological studies of *Bombax malabaricum* flowers. Nat Prod Res 25(2):141–151

Faizi S, Ali M (1999) Shamimin: a new flavonol C-glycoside from leaves of *Bombax ceiba*. Planta Med 65(4):383–385

Fan P, Hay A-E, Marston A, Hostettmann K (2008) Acetylcholinesterase-inhibitory activity of Linarin from *Buddleja davidii*. Structure-activity relationships of related flavonoids, and chemical investigation of *Buddleja nitida*. Pharmaceutical Biol 46(9):596–601

Fernández S, Wasowski C, Paladini AC, Marder M (2004) Sedative and sleep-enhancing properties of linarin, a flavonoid-isolated from *Valeriana officinalis*. Pharmacol Biochem Behav 77(2):399–404

Field FJ, Born E, Mathur SN (1997) Effect of micellar beta-sitosterol on cholesterol metabolism in CaCo-2 cells. J Lipid Res 38:348–360

Fuhrman B, Lavy A, Aviram M (1995) Consumption of red wine with meals reduces the susceptibility of human plasma and low-density lipoproteins to lipid peroxidation. Am Soc Clin Nutr 61:549–554

Ghate VS, Agte VV, Vartak VD (1988) Promising economic potential of Shemul *Bombax ceiba* L. as a tuber crop. Indian J Forestry 11:158–159

Ghose TP (1935) *Bombax malabaricum*. Indian Forester 61:93–103

Gopal H, Gupta RK (1972) Chemical constituents of *Salmalia malabarica* Schott & Endl. Flowers. J Pharma Sci 61:807–808

Grael CFF, Albuquerque S, Lopes JLC (2005) Chemical constituents of Lychnophora pohlii and trypanocidal activity of crude plant extracts and of isolated compounds. Fitoterapia 76:73–82

Habtemariam S (2011) α-glucosidase inhibitory activity of kaempferol-3-O-rutinoside. Nat Prod Commun 6(2):201–203

Hibasami H, Saitoh K, Katsuzaki H, Imai K, Aratanechemuge Y, Komiya T (2004) 2-O-methylisohemigossylic acid lactone, a sesquiterpene, isolated from roots of mokumen (*Gossampinus malabarica*) induces cell death and morphological change indicative of apoptotic chromatin condensation in human promyelotic leukemia HL-60 cells. Int J Mol Med 14(6):1029–1033

Hoffmann-Bohm K, Lotter H, Seligmann O, Wagner H (1992) Antihepatotoxic C-glycosylflav-ones from the leaves of *Allophyllus edulis* var. *edulis* and *gracilis*. Planta Med 58:544–548

Hossain E, Mandal SC, Gupta JK (2011) Phytochemical Screening and in vivo antipyretic activity of the methanol leaf-extract of *Bombax malabaricum* DC (Bombacaceae). Trop J Pharm Res 10(1):55–60

Igarashi K, Ohmuna M (1995) Effect of Isoharnneti, Rhamnetin and Quercetin on the concentrations of cholesterol and lipoperoxide in the serum and liver and on the blood and liver antioxidative enzyme activities in rats. Biosci Biotech Biochem 59:595–601

Itoigawa M, Takeya K, Ito C, FuruKawa H (1999) Structure activity relationship of cardiotonic flavonoids in guinea pig papillary muscle. J Ethnopharmacol 65:267–272

Ivorra MD, Docon MP, Paya M, Villar A (1988) Antihyperglycemic and insulin-releasing effects of beta-sitosterol-3-beta-D-glucoside and its aglycone, beta-sitosterol. Arch Int Pharmacodyn Ther 296:224–231

Izzo AA, Carlo GD, Mascolo N, Capasso F, Autore G (1994) Effects of quercetin on gastrointestinal tract: further studies. Phytother Res 8:179–185

Jahaniani F, Ebrahimi SA, Rahbar-Roshandel N, Mahmoudian M (2005) Xanthomicrol is the main cytotoxic component of *Dracocephalum kotschyii* and a potential anti-cancer agent. Phytochemistry 66(13):1581–1592

Jain V, Verma SK, Katewa SS, Anandjiwala S, Singh B, 4 (2011) Free radical scavenging property of *Bombax ceiba* L. root. Res J Med Plant 5(4):462–470

Jeong H-J, Choi Y, Kim K-Y, Kim M-H, Kim H-M (2011) C-Kit binding properties of Hesperidin (a Major Component of KMP6) as a potential anti-allergic agent. PLoS ONE 6(4):19528. doi:10.1371/journal.pone.0019528

Kapoor VP, Khan PSH, Raina RM, Farooqi MIH (1975) Chemical analysis of seeds: Part III from 40 non-leguminous species. Sci Cult 41:336–339

Kim MJ, Seong AR, Yoo JY, Jin CH, Lee YH, Kim YJ, Lee J, Jun WJ, Yoon HG (2011a) Gallic acid, a histone acetyltransferase inhibitor, suppresses β-amyloid neurotoxicity by inhibiting microglial-mediated neuroinflammation. Mol Nutr Food Res. doi:10.1002/mnfr.201100262

Kim SJ, Chung WS, Kim SS, Ko SG, Um JY (2011b) Antiinflammatory effect of *Oldenlandia diffusa* and its constituent, hentriacontane, through suppression of caspase-1 activation in mouse peritoneal macrophages. Phytother Res 25(10):1537–1546

Koo H, Pearson SK, Scott-Anne K, Abranches J, Cury JA, Rosalen PL, Park YK, Marquis RE, Bowen WH (2002) Effects of apigenin and tt-farnesol on glucosyltransferase activity, biofilm viability and caries development in rats. Oral Microbiol Immunol 17(6):337–343

Kroes BH, van den Berg AJ, Quarles van Ufford HC, van Dijk H, Labadie RP (1992) Anti-inflammatory activity of gallic acid. Planta Med 58(6):499–504

Krogh R, Yunes RA, Andricopulo AD (2000) Structure–activity relationships for the analgesic activity of gallic acid derivatives. Il Farmaco 55(11–12):730–735

Lagnika L, Wenigerb B, Vonthron-Senecheaub C, Sanni A (2009) Antiprotozoal activities of compounds isolated from *Croton lobatus* L. Afr J Infect Dis 3(1):1–5

Lee J-H, Lee JY, Park JH, Jung HS, Kim JS, Kang SS, Kim YS, Han Y (2007) Immunoregulatory activity by daucosterol, a β-sitosterol glycoside, induces protective Th1 immune response against disseminated Candidiasis in mice. Vaccine 25(19):3834–3840. doi:10.1016/j.vaccine.2007.01.108

Lee KM, Lee KW, Jung SK, Lee EJ, Heo YS, Bode AM, Lubet RA, Lee HJ, Dong Z (2010) Kaempferol inhibits UVB-induced COX-2 expression by suppressing Src kinase activity. Biochem Pharmacol 80(12):2042–2049

Liu J (1995) Pharmacology of oleanolic acid and ursolic acid. J Ethnopharmacol 49(2):57–68

Liu Z, Schwimer J, Liu D, Lewis J, Greenway FL, York DA, Woltering EA (2006) Gallic acid is partially responsible for the antiangiogenic activities of *Rubus* leaf extract. Phytother Res 20(9):806–813

Loizou S, Lekakis I, Chrousos GP, Moutsatsou P (2010) Beta-sitosterol exhibits anti-inflammatory activity in human aortic endothelial cells. Mol Nutr Food Res 54(4):551–558

Lorget F, Prouillet C, Petit J-P, Fardelonne P, Brazier M, Wattel A, Kamel S, Mentaverri R (2003) Potent inhibitory effect of naturally occurring flavonoids quercetin and kaempferol on in vitro osteoclastic bone resorption. Biochem Pharmacol 65(1):35–42

Marrassini C, Davicino R, Acevedo C, Anesini C, Gorzalczany S, Ferraro G (2011) Vicenin-2, a potential anti-inflammatory constituent of *Urtica circularis*. J Nat Prod 74:1503–1507

Martínez-González J, Rodríguez-Rodríguez R, González-Díez M, Rodríguez C, Herrera MD, Ruiz-Gutierrez V, Badimon L (2008) Oleanolic acid induces prostacyclin release in human vascular smooth muscle cells through a cyclooxygenase-2-dependent mechanism. J Nutr 138:443–448

McAnlis GT, Mc Enany J, Pearce J, Young IY (1997) The effect of various dietary flavonoids on the susceptibility of low-density lipoproteins to oxidation in vitro using both metallic and nonmetallic oxidising agents. Biochem Soc Trans 25:142–148

Meckes M, Calzada F, Paz D, Rodríguez J, Ponce-Monter H (2002) Inhibitory effect of xanthomicrol and 3α-angeloyloxy-2α-hydroxy-13,14Z-dehydrocativic acid from *Brickellia paniculata* on the contractility of guinea-pig Ileum. Planta Med 68(5):467–469

Min YS, Yim SH, Bai KL, Choi HJ, Jeong JH, Song HJ, Park SY, Ham I, Whang WK, Sohn UD (2005) The effects of apigenin-7-O-beta-D-glucuronopyranoside on reflux oesophagitis and gastritis in rats. Auton Autacoid Pharmacol 25(3):85–91

Monforte MT, Trovato A, Kirjavainen S, Forestieri AM, Galati EM, Lo Curto RB (1995) Biological effects of hesperidin, a *Citrus* flavonoid. (note II): hypolipidemic activity on experimental hypercholesterolemia in rat. Farmaco 50(9):595–599

Mukherjee J, Roy B (1971) Chemical examination of *Salmalia malabarica* Schott & Endl syn *Bombax malabaricum* DC. J Indian Chem Soc 48:769–770

Muruganandan S, Srinivasan K, Gupta S, Gupta PK, Lal J (2005) Effect of mangiferin on hyperglycemia and atherogenicity in streptozotocin diabetic rats. J Ethnopharmacol 97:497–501

Nagaprashantha LD, Vatsyayan R, Singhal J, Fast S, Roby R, Awasthi S, Singhal SS (2011) Anti-cancer effects of novel flavonoid vicenin-2 as a single agent and in synergistic combination with docetaxel in prostate cancer. Biochem Pharmacol 82(9):1100–1109

Nagarathnam D, Cushman M (1991) A short and facile synthetic route to hydroxylated flavones. New syntheses of apigenin, tricin, and luteolin. J Org Chem 56:4884–4887

Niranjan GS, Gupta PC (1973) Anthocyanins from flowers of *Bombax malabaricum*. Planta Med 24(2):196–199

Okada S, Saito M, Kazuyama E, Hanada T, Kawaba Y, Hayashi A, Satoh K, Kanzaki S (2008) Effects of N-hexacosanol on nitric oxide synthase system in diabetic rat nephropathy. Mol Cell Biochem 315(1–2):169–177

Okoko T, Oruambo IF (2009) Inhibitory activity of quercetin and its metabolite on lipopolysaccharide-induced activation of macrophage U937 cells. Food Chem Toxicol 47(4):809–812

Online document 1. Chapter 8. Antibacterial, antioxidant and cytotoxic activity of taraxerol, a petacyclic triterpenoid, isolated from *Pteleopsis myrtifolia* leaves. http://upetd.up.ac.za/thesis/available/etd-05082008-131034/unrestricted/08chapter8.pdf. Accessed 4 Nov 2011

Puckhaber LS, Stipanovic RD (2001) Revised structure for a Sesquiterpene lactone from *Bombax malabaricum*. J Nat Prod 64(2):260–261

Qi Y, Guo S, Xia Z, Xie D (1996) Chemical constituents of *Gossampinus malabarica* (L.) Merr. (II). Zhongguo Zhong Yao Za Zhi 21(4):234–235, 256

Rajnarayana K, Reddy MS, Chaluvadi MR, Krishna DR (2001) Bioflavonoids classification, pharmacological, biochemical effects and therapeutic potential. Indian J Pharmacol 33:2–16

Rajput MS, Mathur V, Agrawal P, Chandrawanshi HK, Pilaniya U (2011) Fibrinolytic activity of kaempferol isolated from the fruits of *Lagenaria siceraria* (Molina) Standley. Nat Prod Res. [Epub ahead of print]

Rastogi RP, Mehrotra BN (1995) Compendium of Indian medicinal plants, vol 4. PID, New Delhi

Reddy MVB, Reddy MK, Duvvuru G, Murthy MM, Caux C, Bodo B (2003) A new Sesquiterpene lactone from *Bombax malabaricum*. Chem Pharm Bull 51(4):458–459

Rizvi SAI, Saxena OC (1974) New glycosides, terpenoids, colouring matters, sugars and fatty compounds from the flowers of *Salmalia malabarica*. Arzneim-Forsch 24(3):285–287

Robak J, Glyglewski RJ (1988) Flavonoids are scavengers of superoxide anions. Biochem Pharmacol 37:837–841

Said A, Aboutable EA, Nofal SM, Tokuda H, Raslan M (2011) Phytoconstituents and bioactivity evaluation of *Bombax ceiba* L. flowers. J Trad Med 28:55–62

Saleem R, Ahmad M, Hussain SA, Qazi AM, Ahmad SI, Qazi MH, Ali M, Faizi S, Akhtar S, Hussain SN (1999) Hypotensive, hypoglycaemic and toxicological studies on the Flavonol C-glucoside Shamimin from *Bombax ceiba*. Planta Med 65:331–334

Saleem R, Ahmad SI, Ahmad M, Faizi Z, Rehman S, Ali M, Faizi S (2003) Hypotensive activity and toxicology of constituents from *Bombax ceiba* stem bark. Biol Pharm Bull 26(1):41–46

Sanchez De RVR, Somoza B, Ortega T, Villar A (1996) Flavonoids action on heart. Planta Med 62:554–562

Sankaram AVB, Reddy NS, Shoolery JN (1981) New sesquiterpenoids of *Bombax malabaricum*. Phytochemistry 20:1877–1881

Sengupta S, Mukherjee A, Goswami R, Basu S (2009) Hypoglycemic activity of the antioxidant saponarin, characterized as alpha-glucosidase inhibitor present in *Tinospora cordifolia*. J Enzyme Inhib Med Chem 24(3):684–690

Seshadri V, Batta AK, Rangaswami S (1971) Phenolic components of *Bombax malabaricum* (Root-Bark). Curr Sci 23:630

Shahat AA, Hassan RA, Nazif NM, Van Miert S, Pieters L, Hammuda FM, Vlietinck AJ (2003) Isolation of mangiferin from *Bombax malabaricum* and structure revision of shamimin. Planta Med 69(11):1068–1070

Shinbori C, Saito M, Kinoshita Y, Satoh I, Kono T, Hanada T, Nanba E, Adachi K, Suzuki H, Yamada M, Satoh K (2006) N-hexacosanol reverses diabetic induced muscarinic hypercontractility of ileum in the rat. European J Pharmacol 545(2–3):177–184

Shinha H, Ki-hyun S, Dongsool Y, Sookyeon L, Chong-kil L, Nam-ju H, Kyungjae K (2002) The effect of linarin on LPS-induced cytokine production and nitric oxide inhibition in murine macrophages cell line RAW264.7. Arch Pharm Res 25(2):170–177

Singh AB, Yadav DK, Maurya R, Srivastava AK (2009) Antihyperglycaemic activity of alpha-amyrin acetate in rats and db/db mice. Nat Prod Res 23(9):876–882

Sood RP, Suri KA, Suri OP, Dhar KL, Atal CK (1982) Sesquiterpene lactone from *Salmalia malabarica*. Phytochemistry 21(8):2125–2126

Sreeramulu K, Rao KV, Venkatarao C, Gunasekar D (2001) A new naphthoquinone from *Bombax malabaricum*. J Asian Nat Prod Res 3:261–265

Stoilova I, Jirovetz L, Stoyanova A, Krastanov A, Gargova S, Ho L (2008) Antioxidant activity of the polyphenol mangiferin. EJEAFChe 7(13):2706–2716

Subarnas A, Tadano T, Nakahata N, Arai Y, Kinemuchi H, Oshima Y, Kisara K, Ohizumi Y (1993) A possible mechanism of antidepressant activity of beta-amyrin palmitate isolated from Lobelia inflata leaves in the forced swimming test. Life Sci 52(3):289–296

Szakiel A, Ruszkowski D, Grudniak A, Kurek A, Wolska KI, Doligalska M, Janiszowska W (2008) Antibacterial and antiparasitic activity of oleanolic acid and its glycosides isolated from marigold (*Calendula officinalis*). Planta Med 74(14):1709–1715

Tang R, Chen K, Cosentino M, Lee KH (1994) Apigenin-7-O-β-D-glucopyranoside, an anti-HIV principle from *Kummerowia striata* Bioorganic & Medicinal Chemistry Letters 4(3):455–458

Velozo LSM, Ferreira MJP, Santos MIS, Moreira DL, Guimaraes EF, Emerenciano VP, Kaplan MAC (2009) C-glycosyl flavones from *Peperomia blanda*. Fitoterapia 80:119–122

Villaseñor IM, Angelada J, Canlas AP, Echegoyen D (2002) Bioactivity studies on beta-sitosterol and its glucoside. Phytother Res 16(5):417–421

Vitcheva V, Simeonova R, Krasteva I, Yotova M, Nikolov S, Mitcheva M (2011) Hepatoprotective effects of saponarin, isolated from *Gypsophila trichotoma* Wend. on cocaine-induced oxidative stress in rats. Redox Rep 16(2):56–61

Wal P, Wal A, Sharma G, Rai AK (2011) Biological activities of lupeol. Syst Rev Pharm 2:96–103

Wang H, Zeng Z, Zeng H-P (2003) Study on chemical constituents of petroleum ether fraction of alcohol extract from the flower of *Bombax malabaricum*. Chem Ind Forest Products. doi:cnki:issn:0253-2417.0.2003-01-018

Wang R-R, Gao Y-D, Ma C-H, Zhang X-J, Huang C-G, Huang J-F, Zheng Y-T (2011) Mangiferin, an Anti-HIV-1 agent targeting protease and effective against resistant strains. Molecules 16:4264–4277

Wu J, Zhang XH, Zhang SW, Xuan LJ (2008) Three novel compounds from the flowers of *Bombax malabaricum*. Helvetica Chimica Acta 91(1):136–143

Yin MC, Chan KC (2007) Nonenzymatic antioxidative and anti-glycative effects of oleanolic acid and ursolic acid. J Agric Food Chem 55(17):7177–7181

You YJ, Nam NH, Kim HM, Bae KH, Ahn BZ (2003) Antiangiogenic activity of lupeol from *Bombax ceiba*. Phytother Res 17(4):341–344

Zhang X, Zhu H, Zhang S, Yu Q, Xuan L (2007) Sesquiterpenoids from *Bombax malabaricum*. J Nat Prod 70:1526–1528

Zoechling WA, Reiter E, Eder R, Wendelin S, Liebner F, Jungbauer A (2009) The flavonoid kaempferol is responsible for the majority of estrogenic activity in Red Wine. AJEV 60(2):223–232

Chapter 4
Pharmacological Investigations and Toxicity Studies

Abstract Red Silk-Cotton tree has been scrutinized for its pharmacology in various parts of the world. Different parts of *Bombax ceiba* has shown to possess many biological properties predominantly antioxidant, antimicrobial, anti-inflammatory, analgesic, anabolic, hepatoprotective, hypotensive and hypoglycemic activities. It has proved to be safe in various toxicity studies. However, it still needs extensive scientific exploration.

Keywords Antioxidant · Antimicrobial · Antitumor · Anabolic · Hepatoprotective · Hypotensive · Hypoglycemic

4.1 Introduction

In the last decade, *Bombax ceiba* has attracted scientific attention that resulted in exploration of many novel chemical compounds as well as validation of its traditional uses in many diseases of man. Animal experimental studies have shown that *B. ceiba* has important pharmacological activities without much toxicity. Moreover, human studies have further strengthened its therapeutic role as an anabolic, antihyperglycemic, hypolipidemic and fibrinolysis-enhancing agent. The following discussion will restrict to its important pharmacological activities with passing comments on its other effects.

V. Jain and S. K. Verma, *Pharmacology of Bombax ceiba Linn.*,
SpringerBriefs in Pharmacology and Toxicology,
DOI: 10.1007/978-3-642-27904-1_4, © The Author(s) 2012

4.2 Antioxidant Activity

Methanolic extract of whole plant material of *B. ceiba* showed DPPH radical scavenging activity with an IC_{50} of 68 µg/ml among 26 folk herbal medicinal plant extracts popularly used in Taiwan. *Bombax ceiba* did show the higher scavenging activity then that of *Ginkgo biloba* leaf extract (IC_{50} 930 µg/ml) used as reference. It also exhibited significant protection on $\Psi \times 174$ supercoiled DNA against strand cleavage induced by UV irradiated H_2O_2. The magnitude of inhibition of open circular DNA formation with *B. ceiba* was 0.35 which is quite comparable to that of catechin 0.38, used as a reference control (Shyur et al. 2005). This effective prevention of hydroxyl radical-induced DNA damage by *B. ceiba* needs further attention.

Dar et al. (2005) studied mangiferin, a xanthone isolated from methanolic extract of fresh leaves of *B. ceiba* and tested it (1) and its acetyl (1a), cinnamoyl (1b) and methyl (2) derivatives along with the methanolic extract (BCL) and the filtrate (BCM) for antioxidant activity in DPPH free radical scavenging assay, Deoxyribose degradation assay and non-enzymatic lipid peroxidation in liposomes. Methanolic extract along with mangiferin showed DPPH scavenging activity where mangiferin has shown IC_{50} value of 5.8 ± 0.96 µg/ml which was quite comparable to IC_{50} of Rutin (5.56 ± 0.33 µg/ml). Mangiferin appears to be a better antioxidant compound than 1a and 1b which bears no 6,7-dihydroxylated structure (Catechol moiety). However, methyl derivative of mangiferin (2) was devoid of DPPH radical scavenging activity even at concentration of 200 µg/ml suggesting that the presence of methoxy group abolishes the antioxidant activity. Furthermore, mangiferin could not significantly protect against either deoxyribose damage or lipid peroxidation.

Surveswaran et al. (2007) have studied 133 Indian medicinal plants for their antioxidant activity using three in vitro assays (ABTS, DPPH and FRAP). Eighty percent methanolic extract of gum (*Mochrasa*) of *B. ceiba* was evaluated for its total phenolic contents and antioxidant activity. Total phenolic content of the gum was found to be 5.89 GAE/100 g dry weight. Gum has shown 55.38 mmol TEAC/100 g dry weight in ABTS assay and 80.12 mmol TEAC/100 g DW in DPPH assay, whereas FRAP assay has shown value of 9.06 µmol TEAC/100 g DW.

Lately, Vieira et al. (2009) have reported antioxidant activity of defatted methanolic extract of flowers of *B. ceiba*. The EC_{50} (µg/ml) obtained for DPPH radical was 87 and for lipid peroxidation of rat liver microsomes and soy bean phosphatidylcholine liposomes induced by ascorbyl radicals were 141 and 105, respectively and by peroxynitrite were 115 and 77, respectively. The extract also inhibited myeloperoxidase activity and kinetics of this enzyme inhibition gave $K_{0.5}$ value of 264 µg/ml. Cytotoxicity of the extract was also monitored through the mitochondrial activity in the Vero cell line. Extract started to show toxicity toward cells at 750 µg/ml which was a much higher concentration than those which showed antioxidant activity.

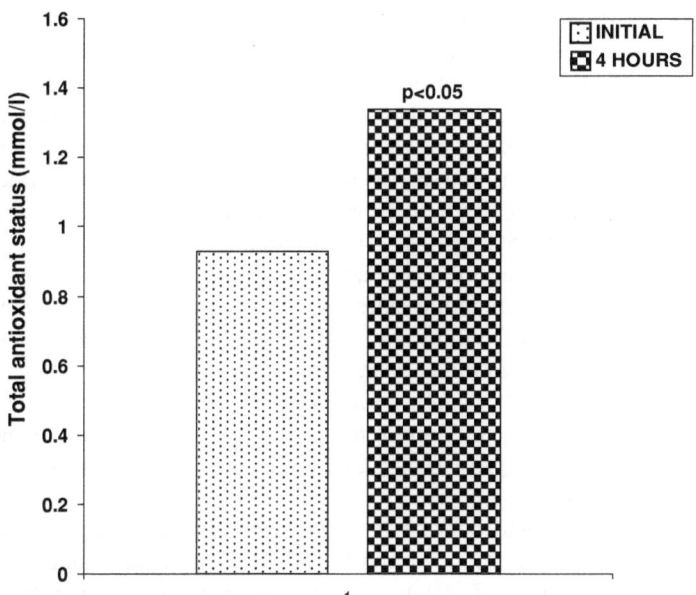

Fig. 4.1 Acute effect of 3 g *Bombax ceiba* root powder administration on total antioxidant status in ten healthy volunteers (Jain et al. 2011a)

Yu et al. (2011) assessed the antioxidant activities of water, 50% ethanol, 80% acetone extracts from flowers of *Bombax malabaricum* on DPPH radical scavenging, ORAC, reducing power and inhibition on phosphatidylcholine liposome peroxidation. All extracts showed remarkable antioxidant capacity compared with ascorbic/gallic acid.

Recently, Jain et al. (2011a) have assessed antioxidant activity of *B. ceiba* root using DPPH radical scavenging and reducing power assay. Methanolic extract of root showed high amounts of phenolics (30.9% w/w) and tannins (15.45% w/w) and a very good DPPH scavenging activity (EC_{50} 15.07 μg/ml) in a dose-dependent manner as well as dose-dependent reduction ability (Fe^{3+} to Fe^{2+} transformation) with a maximum absorbance of 1.11 at a concentration of 500 μg of extract. Furthermore, acute study in human healthy volunteers showed a significant ($p < 0.05$) rise in total antioxidant status at the end of 4 h after administration of 3 g root powder (Fig. 4.1).

4.3 Antimicrobial Activity

The plant is endowed with strong capacity to fight against microorganisms such as bacteria, fungi and viral attacks. Each part possesses antimicrobial activity as has been demonstrated in various in vitro experimental studies. The research can be traced back to 1968 when Dhar et al. demonstrated antiviral activity of 50% ethanolic extract of *B. ceiba* flowers at a maximum tolerated dose of 250 mg/kg body weight against Ranikhet disease virus.

Antimicrobial research work then took a pause of 20 years when in 1991 Mishra et al. reported the effect of water extract of leaves of *B. ceiba* against fungi. They observed that water extract of leaves of *B. ceiba* demonstrated 90.4% mycelial inhibition against *Epidermophyton floccosum*, 80.4% against *Tricophyton mentagrophytes* and 75.25% against *Microsporum gypseum*.

In 1999, Faizi and Ali reported antibacterial and antifungal activity of this plant. Shamimin, isolated from ethanolic extract of its fresh leaves at a concentration of 100 μg, showed zone of inhibition (8–12 mm) against three Gram-positive (*Listeria monocytogenes*, *Bacillus subtilis* and *Streptococcus pyogenes*) and five Gram-negative (*Shigella sonnei*, *Salmonella typhi*, *Enterobacter cloacae*, *Pseudomonas aeruginosa* and *Shigella flexneri*) bacteria. Furthermore, at a similar concentration, it inhibited growth of *Candida albicans* among the five fungi in an in vitro antifungal assay.

Interestingly, leaves and bark of *B. ceiba* were also tested against multi-drug resistant *S. typhi* strains MTCC 531 and B 330. Methanolic and aqueous extract of bark at a concentration of 50 mg/ml showed strong antimicrobial activity (≥5–9 mm diameter of zone of inhibition) against *S. typhi* strain MTCC 531. Methanolic extract of bark showed ≥9–15 mm diameter of zone of inhibition while aqueous extract showed ≥5–9 mm diameter of zone of inhibition against *S. typhi* strain B 330. Minimum inhibitory concentration (MIC) of methanolic bark extract assessed was found to be 256 μg/ml against both the strains used. Methanolic extract of leaves, on the other hand, showed ≥5–9 mm diameter of zone of inhibition against only strain MTCC 530 and aqueous extract was completely ineffective (Rani and Khullar 2004).

Wang and Huang (2005) while screening 50 Taiwanese folk medicinal plants against ten strains of *Helicobacter pylori* observed that ethanolic extract (95%) of *B. ceiba* root has shown anti-*H. pylori* activity against all the strains. MIC values were ranging from 1.28 to 5.12 mg/ml.

Not only the plant part, such as flower, leaves, stem bark and root showed antimicrobial activity, but also the fungal endophytes isolated from leaves and stem have shown anti-*Mycobacterium tuberculosis* activity using microplate Alamar blue assay and anti-Herpes Simplex Virus activity in Vero cell assay. However, no antimalarial activity was observed (Wiyakrutta et al. 2004).

It is interesting to note that as far as antimicrobial activity is concerned, brunt of work is on flower, leaves and stem bark. However, there are very few studies on root. Recently, the methanolic extract of root in five concentrations (50, 25, 12.5, 6.25 and 3.12 mg/ml) has shown a significant dose-dependent antibacterial potential in in vitro agar well diffusion assay. Zone of inhibition (mm diameter) obtained at the highest concentration of methanolic extract (50 mg/ml) for Gram-positive bacteria *Staphylococcus aureus*, *B. subtilis* and Gram-negative bacteria *Escherichia coli* and *Klebsiella pneumoniae* was 17.0, 16.0, 17.16 and 17.10, respectively (Jain et al. 2011b).

Methanol, *n*-hexane, chloroform and carbon tetrachloride extracts of *B. ceiba* root were tested against 13 bacteria and three fungi. The zone of inhibition produced was found to be 10–14, 7–13, 9–15 and 13–20 mm, respectively at a concentration of 200 μg/disc against nine bacterial strains and two fungal strains. The significant activity was found with methanolic extract against *S. typhi* and *Pseudomonas*, hexane extract against *Sarcina lutea* and *Pseudomonas* and chloroform extract against *Vibrio mimicus*. The carbon tetrachloride extract showed prominent (>15 mm) activity against almost all bacterial strains tested. As far as antifungal activity against the three fungi is concerned, the chloroform and carbon tetrachloride extract showed profound activity against *Aspergillus niger* and *C. albicans*, respectively (Islam et al. 2011).

4.4 Anthelmintic Activity

Methanolic extract of *B. malabaricum* leaves was evaluated for its anthelmintic activity against a live trematode *Parmphistomum explanatum* at different doses by comparing with albendazole. All parasites died with the doses selected (10, 25, 50 and 100 mg/ml) within a short period of time (<45 min). At the dose of 100 mg/ml, it showed maximum efficacy, by paralyzing and killing trematodes in 18.50 ± 0.62 and 22.17 ± 0.48 min, respectively. Interestingly, it proved more powerful than albendazole which required more time for paralysis (73.17 ± 1.45 min) or death (82.33 ± 1.38 min) with a level of significance of $p < 0.001$ (Hossain et al. 2011a).

4.5 Larvicidal Activity

Powdered leaves and methanolic extract of *B. malabaricum* were evaluated for their larvicidal activity against first, second, third and fourth in-star larval forms of filarial vector *Culex quinquefasciatus*. At all the concentrations of powdered leaves there was a significant ($p < 0.01$) larval mortality. The mortality was higher in 50 ppm dose of methanolic extract (Hossain et al. 2011b).

4.6 Anti-inflammatory Activity

Lin et al. (1992) have studied effect of stem bark, stem xylem and root of *B. malabaricum* against carrageenan-induced edema in male wistar rats (150–180 g). Control group received normal saline. Other group received indomethacin (10 mg/kg) and test group was treated with aqueous extract 10 mg/kg body weight subcutaneously, followed 1 h after 0.05 ml of carrageenan (1% w/v in saline) into rats left hind paw. Indomethacin showed 51% reduction in edema inhibition after 5 h. Root and stem xylem were proved to be better than the indomethacin by significantly ($p < 0.01$) reducing edema inhibition after 5 h by 79 and 74%, respectively while bark showed 46% inhibition rate.

Ethyl acetate soluble fraction of alcoholic extract from flowers of *B. ceiba* had shown anti-inflammatory activity. At the dose of 200–500 mg/kg i.p. it markedly inhibited hind paw edema induced by injection of fresh egg white or carrageenan in rats or mouse (Xu et al. 1993).

Oral administration of 70% methanolic extract of flowers in a dose of 25 and 50 mg/100 g body weight reduced carrageenan-induced rat hind paw edema by 22.9 and 37% after 4 h (Said et al. 2011).

Dar et al. (2005) have observed that different extracts and mangiferin, isolated from fresh leaves of *B. ceiba* at 100 mg/kg failed to exhibit detectable anti-inflammatory activity when subjected to carrageenan-induced rat paw edema.

Aqueous extract of *B. malabaricum* gum (270 mg/kg) significantly reduced the ulcer score and myeloperoxidase activity in indomethacin and iodoacetamide-induced colitis in Sprague-Dawley rats. In a dose of 500 mg/kg, it also significantly reduced the ulcer score and myeloperoxidase activity in acetic acid-induced colitis in Swiss albino mice. These two extracts i.e. 270 mg/kg in rats and 500 mg/kg in mice were found to be comparable with prednisolone (10 mg/kg) and 5-aminosalicylic acid (100 mg/kg). Aqueous extract reduced edema of the intestinal tissue, provided ulcer protection and lowered myeloperoxidase activity in a dose dependent manner (Jagtap et al. 2011).

4.7 Analgesic Effect

Gum of the plant was reported to have analgesic activity as tested by rat-tail hot wire technique in a preliminary study (Gupta et al. 1981).

Dar et al. (2005) demonstrated that methanolic extract of *B. ceiba* leaves, its fractions and mangiferin induced a significant and dose-dependent analgesic effect in acetic acid writhing and hot plate test in mice. On further elucidating the mechanism of analgesia, it was observed that acetic acid-induced pain was antagonized in the presence of morphine (65%) and mangiferin (70%); however using naloxone, the corresponding effect was only 19 and 28%, respectively suggesting the involvement of opioid pathway. Interestingly, naloxone reversed around 38% nociceptive effect of mangiferin but the effect induced by methanolic extract of its leaves and fractions remains unchanged in the presence of naloxone indicating that their analgesic effect was independent of opioid receptors.

Said et al. (2011) have demonstrated that administration of 25 and 50 mg/100 g body weight of 70% methanolic extract of flowers delayed the mean reaction time on hot plate by 56.5 and 66.2% after 1 h, respectively in mice. In the same doses, 70% methanolic extract reduced the acetic acid-induced writhing score by 27.3 and 47.6% respectively in mice indicating its peripheral analgesic effect.

4.8 Antipyretic Activity

The antipyretic activity of methanolic extract of *B. malabaricum* leaves has been reported for the first time by Hossain et al. (2011c). The methanolic extract showed significant antipyretic activity in the doses of 200 mg/kg ($p < 0.05$) and 400 mg/kg ($p < 0.01$) in Baker's yeast-induced pyrexia and was higher than control. The maximum antipyretic activity for the extract occurred at 6 h while that of paracetamol, the reference standard, occurred at 3 h.

4.9 Anti-angiogenic and Antimutagenic Activities

Lupeol, isolated from methanolic extract of stem bark of *B. ceiba*, exhibited significant antiangiogenic activity on in vitro tube formation of human umbilical venous endothelial cells (HUVEC) with inhibition rate of more than 80% at 50 μg/ml and 40–60% inhibition at a concentration of 30 μg/ml. In contrary, it did not affect the growth of tumor cell lines viz. SK-MEL-2 (human melanoma), A549 (human lung carcinoma) and B16-F10 (murine melanoma) meaning thereby, that it does not have significant cytotoxicity ($ED_{50} > 30$ μg/ml) in SRB assay (You et al. 2003).

Nam et al. (2003) have screened 58 Vietnamese medicinal plants for the in vitro angiogenesis inhibitory activity using HUVEC. Out of 58, seven plants exhibited strong to moderate inhibitory activity on tube-like formation induced by HUVEC in the in vitro angiogenesis assay. Methanolic extract of *B. ceiba* stem (3 μg/ml) did not show inhibitory activity in the in vitro angiogenesis assay, on the other hand, stem and leaves of *Ceiba pentandra*—the White Silk-Cotton tree (Family-Bombacaceae) showed strong antiangiogenic activity.

Methanolic extract of *B. ceiba* has also exhibited some antimutagenicity against the heterocyclic amine Trp-P-1 using *Salmonella typhimurium* strain TA98 among 108 edible Thai plant species evaluated (Nakahara et al. 2002). Lupeol isolated from Mokumen (*B. ceiba*) has induced apoptosis on DNA human promyelotic HL-60 leukemic cells (Aratanechemuge et al. 2004). Qi et al. (2008) have also shown significant antitumor activity of roots of *B. ceiba* on ascites sarcoma-180 (S-180) and U14-transplanted mice in a dose-dependent manner.

Chen et al. (2009) have also studied effect of flavonoids isolated from *B. ceiba* flowers on inhibition of fatty acid synthase (FAS) enzyme. FAS has been found to be overexpressed in various types of cancers. Pharmacological inhibition of FAS can repress cancer cell proliferation and thus FAS is an excellent drug target for cancer therapy. However MIC with *B. ceiba* flavonoid was obtained as 247.98 μg/ml which was quite high as compared to flavanoids isolated from other plants.

Petroleum ether, chloroform, ethyl acetate, butanol and 70% methanolic extract of flowers were evaluated for their antitumor activity. All the extracts significantly inhibited Epstein Barr Virus, early antigen induced by TPA in Raji cells and ethyl acetate extract exhibited the maximum inhibition response (Said et al. 2011).

4.10 Hepatoprotective Activity

Aqueous extract of *B. malabaricum* bark (concentrated into 1 g/ml) has been evaluated for its effect on CCL_4-induced hepatotoxicity in male wistar albino rats. Control group received 3 ml/kg olive oil intraperitoneally while other group received CCL_4/Olive oil (1:1 V/V) subcutaneously in same dose. Test group received aqueous extract (1 g/kg of body weight) intraperitoneally. There was a significant ($p < 0.0001$) reduction of 51% in CCL_4-induced AST and 71% reduction in ALT in rats administered aqueous extract. It also demonstrated a significant hepatoprotective effect on CCL_4-induced liver fatty degeneration, diffused sinusoid enlargement and cell necrosis as compared to other groups (Chiu et al. 1992).

Lin et al. (1992) have evaluated aqueous extract of stem bark, stem xylem and roots of *B. malabaricum* for their effect on CCL_4-induced hepatotoxicity in male wistar rats. A significant ($p < 0.01$) decrease in CCL_4-induced AST and ALT enzyme levels was obtained in test group showing that bark and root of the plant has the best protective capacity against CCL_4-induced hepatotoxicity which was also shown histopathologically. Surprisingly, treatment with all three extracts of stem bark, stem xylem and roots of *C. pentandra* were not effective in protecting liver against CCL_4-induced hepatic damage.

Dar et al. (2005) have evaluated mangiferin, methanolic extract of *B. ceiba* leaves and its fraction on CCL_4-induced liver injury test in Wistar rats. Administration of CCL_4 significantly raised serum AST and ALT levels in control group. Pretreatment of animals with methanolic extract and its fraction did not cause any change in the enzyme levels whereas, after treatment with mangiferin (0.1,1 and 10 mg/kg), both the enzymes showed dose-dependent decline of about 34, 47 and 62%, respectively which were significantly lower than CCL_4-treated animals. It appears that hepatoprotective activity of mangiferin is more likely to be mediated due to its inherent free radical scavenging action.

Methanolic extract of flowers of *B. ceiba* in a dose of 150, 300 and 450 mg/kg i.p. was investigated against hepatotoxicity produced by combination of two antitubercular drugs, isoniazid and rifampicin in rats. At all the doses, there was a significant decrease in AST, ALT, ALP and total Bilirubin levels along with an increase in total protein levels. It also significantly decreased the level of TBARS and elevated the level of GSH. Histology of the liver section of the animals treated with methanolic extract also improved the hepatotoxicity caused by antitubercular drugs (Ravi et al. 2010).

Recently, Said et al. (2011) have shown that oral administration of 70% methanolic extract of flowers of *B. ceiba* in the doses of 250 and 500 mg/kg body weight significantly reduced paracetamol-induced hepatic enzyme levels in rats. At these concentrations, the extract reduced ALT levels by 26.4 and 27.8% and AST levels was reduced by 13.7 and 17.0%, respectively showing a significant hepatoprotective nature of its flowers.

4.11 Hypotensive Activity

Saleem et al. (1999) had first time evaluated hypotensive activity of methanolic and aqueous extract of *B. ceiba* leaves in Sprague-Dawley rats. Methanolic extract and residue fraction of methanolic extract reduced arterial blood pressure by 45–55% at a dose of 70 mg/kg and the effect lasted for 1 h. Aqueous extract also dropped blood pressure by 51% at a dose of 30 mg/kg and the effect lasted for 2–4 min. Novel compound Shamimin, isolated from its leaves (Faizi and Ali 1999), also showed significant potency as a hypotensive agent at doses of 15, 3 and 1 mg/kg with 81, 67 and 51% fall in blood pressure. At lower doses (1 and 3 mg/kg) the effect was brief and returned to normal within 1 min while effect lasted for 2–4 min at the dose of 15 mg/kg. The effect of all these extracts was nullified by prior treatment with atropine. However, later Shahat et al. (2003) reported that Shamimin isolated from the leaves is in fact mangiferin.

Saleem et al. (2003) had also first time evaluated *B. ceiba* stem bark, flowers and fruit pulp for hypotensive activity in Sprague-Dawley rats. I.V. administration of petroleum ether fraction of stem bark (BCBP) at a dose of 10 mg/kg reduced mean arterial blood pressure (MABP) by 58% while lupeol at a dose of 5 and 15 mg/kg reduced MABP by 44 and 52%, respectively. Methanolic extract of defatted stem bark (BCBM) reduced MABP by 30 and 51% at doses of 10 and 30 mg/kg, respectively. Soluble filtrate (BCBMM) obtained from fraction BCBM was found to have more hypotensive activity. It reduced MABP by 31 and 65% at the doses of 3 and 15 mg/kg respectively and the effect lasted for 4 min. Shamimicin isolated from BCBMM, however, did not reduce MABP significantly at the dose of 15 mg/kg. Methanolic extract of fresh flowers and fruit pulp were found as the most active hypotensive agents showing more than 50% reduction in MABP at a dose of 30 mg/kg and the effect lasted for 1–3 min. The most active fraction of stem bark (BCBMM) when administered orally (200 mg/kg/d) for 5 days to Sprague-Dawley rats, also significantly ($p < 0.01$) reduced MABP by 13.2%. This low percentage of hypotension as compared to intravenous administration indicates that active ingredient of BCBMM has low absorbance in blood either due to its very hydrophilic nature or biotransformation.

4.12 Hypoglycemic Activity

Fifty percent ethanolic extract of *B. ceiba* stem bark exhibited hypoglycemic activity in albino rats at a maximum tolerated dose of 50 mg/Kg body weight. Similarly 50% ethanolic extract of *B. ceiba* flowers also exhibited hypoglycemic activity at maximum tolerated dose of 250 mg/Kg (Dhar et al. 1968).

Hypoglycemic effect of Shamimin (mangiferin) isolated from methanolic extract of *B. ceiba* leaves was also evaluated in Sprague-Dawley rats. Intraperitoneal administration of Shamimin produced significant reduction in fasting blood sugar (15%)

at a dose of 500 mg/Kg after 1 h of administration which further reduced blood glucose by 26.6% at the end of 6 h. At a dose of 50 mg/kg it did not produce significant change in blood sugar levels. However, after 24 h all the animals, who were administered with 500 mg/kg of Shamimin died (Saleem et al. 1999).

Root powder of *B. ceiba* has been evaluated by Verma et al. (2008) for its hypoglycemic activity for the first time in type 2 diabetic individuals in a single-blinded, placebo-controlled study. Group I constituted type 2 diabetics uncontrolled on OHA, while group II were also type 2 diabetics, not receiving any antidiabetic drug and were uncontrolled on dietary restrictions and exercise program. In group I, OHA was supplemented with 1.5 g of *B. ceiba* root powder while in group II, 3 g root powder was given as sole therapy in two divided doses for 8 weeks. There was significant ($p < 0.02$) reduction in the fasting blood glucose levels in both the groups at the end of 8 weeks of *B. ceiba* administration. However, the antihyperglycemic effect was more pronounced when it was supplemented to conventional oral hypoglycemic regimen, than when it was the sole therapeutic agent (30% in Group I versus 19% in Group II). It was tolerated well without any adverse reaction and drug interaction (Fig. 4.2).

4.13 Fibrinolysis-Enhancing Activity

An acute dose-dependent response of *B. ceiba* root powder on fibrinolytic activity (FA) has been observed in healthy human volunteers (Verma et al. 2006). There was progressive rise in FA from 0.75 to 10 g doses while 1.5 g was found to be the minimum effective dose (Fig. 4.3). Furthermore, long-term administration of root powder in a dose of 1.5 g for 1 month has also exhibited a significant ($p < 0.05$) increase in FA by 33% and reduction in fibrinogen level. Enhancement in FA was also observed in type 2 diabetic and ischemic heart disease individuals after administration of 3 g *B. ceiba* root powder for 12 weeks (Jain 2009).

4.14 Androgenic and Anabolic Activity

Young roots of *B. ceiba* also known as *Semal-musli* are used traditionally in Indian subcontinent as aphrodisiac. Its juice is considered nutritive, restorative and sexual stimulant. Bhargava et al. (2011) studied lyophilized aqueous extract of roots on sexual behavior, spermatogenesis and anabolic effects in male albino rats in presence of female rats. A gain in body weight was achieved. There was significant ($p < 0.05$) improvement in mount, intromission and ejaculation frequencies. Seminal fructose content and epididymal sperm counts were also significantly improved. Penile erection index was also higher compared to control group.

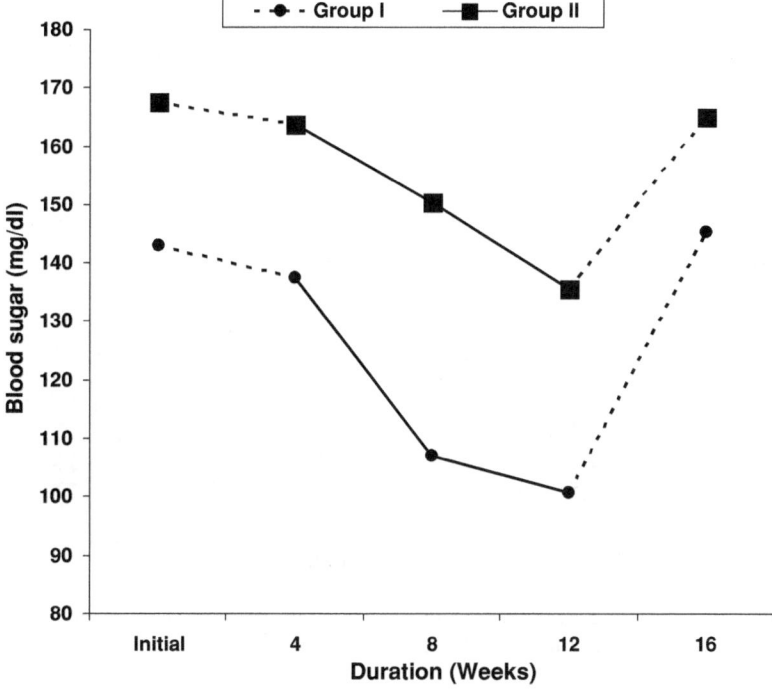

Fig. 4.2 Effect of *Bombax ceiba* root powder on fasting blood sugar levels (mg/dl) in type 2 diabetic individuals. (*Dotted lines*—Placebo administration; *Full lines*: *B. ceiba* root powder administration). Group I: Type-2 diabetics uncontrolled on OHA. Group II: Type-2 diabetics not on OHA (Verma et al. 2008)

Recently, anabolic effect of *B. ceiba* root powder in a patient of involuntary weight loss without any detectable cause has been demonstrated by Verma et al. (2011). Administration of 1.5 g *B. ceiba* root powder with milk for 24 weeks led to a progressive weight gain and attainment of 51 kg with achievement of normal body mass index of 19.9 kg/m^2 at the end of 6 months (Fig. 4.4). Moreover, along with the weight gain, there was 147% rise in FA and marked improvement in the total antioxidant status without any undesirable side effects (Fig. 4.5).

4.15 Other Biological Activities

Flowers of *B. ceiba* are a part of herbal remedy (Five-flower tea) used to alleviate symptoms of Hot Qi in children by Chinese parents. Out of 1,060 Chinese parents interviewed, 57.6% used the five flower tea as a remedy for the symptoms of Hot Qi such as eye discharge, sore throat, halitosis, constipation and irritability (Kong et al. 2006).

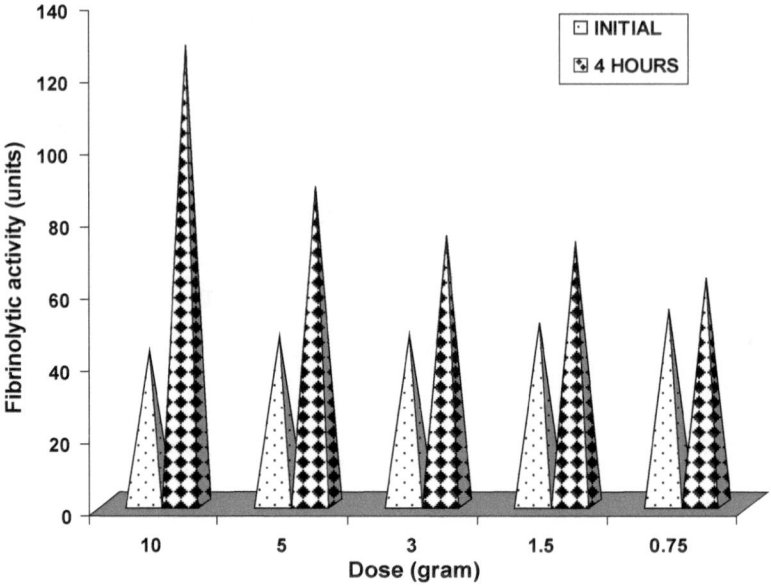

Fig. 4.3 Acute effect of *Bombax ceiba* root powder in different doses on fibrinolytic activity in five healthy individuals (Verma et al. 2006)

The seed extract exhibited hemolytic activity against human ABO red blood cells (Chowdhuri and Chatterjee 1973). Leaves of the seedlings exhibited cholinesterase enzyme activity (Gupta and Gupta 1997). The water, ethanolic and acetone extracts of bark exhibited angiotensin converting enzyme (ACE) inhibitory activity in vitro (Somanadham et al. 1999).

The aqueous extract of its seeds has moderate oxytocic activity on rat, guinea pig and rabbit's uterine strips (Misra et al. 1968). It is partly direct and mainly indirect stimulant to guinea pig ileum and myocardiac stimulant to frog's heart in situ (Misra and Mishra 1966). The ethanolic extract of seeds showed spasmolytic activity on isolated strips of rabbit jejunum and dog ileum in situ and abolished the spasm induced by acetylcholine but had no effect on histamine and barium chloride-induced spasms. It also stimulated respiration of dogs (Sharma and Mishra 1969). The fruit extract exhibited mild activity on the non-gravid adult female albino rats (Dhawan and Saxena 1958).

Aqueous extract of flowers of *B. ceiba* has also demonstrated cardioprotective effect against adriamycin-induced myocardial infarction in rats along with significant free radical scavenging activity (Patel et al. 2011). Moreover, it has been observed that root powder of this plant also modifies significantly many of the coronary risk factors such as atherogenic lipids, fibrinogen and oxidative stress in patients with ischemic heart disease (Jain 2009).

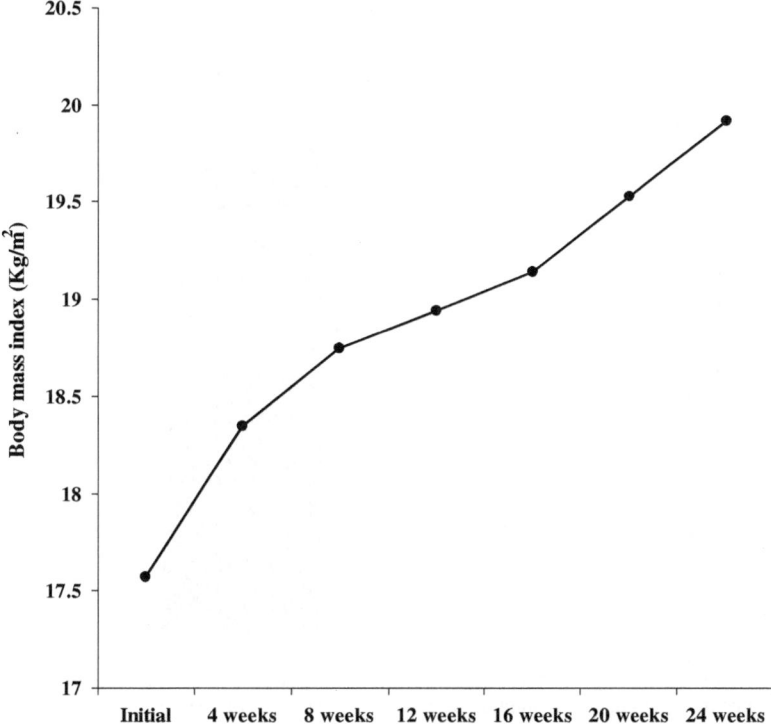

Fig. 4.4 Progressive increase in Body Mass Index (BMI) in a patient of involuntary weight loss (Verma et al. 2011)

4.16 Toxicity Studies

All the parts of the plant have been used for centuries by traditional people for various ailments without any significant adverse effects and toxicity. Recently, higher doses for acute study and 3 g root powder administration for 12 weeks have not resulted in any untoward effect (Jain 2009). However, isolated components from leaves and stem have been evaluated for acute toxicity in animals.

Aqueous, methanolic and residue fraction of methanolic extract along with Shamimin isolated from *B. ceiba* leaves were evaluated for acute toxicity in NMR-1 mice. Aqueous and methanolic extract were orally administered daily for 7 days and also subjected to tissue analysis. Mice showed a high tolerance to all the test substances and no mortality was observed even at a high dose of 1 g/kg. There was no change observed in general behavior and other physiological activities of the animals; however, residue fraction of methanolic extract caused diuresis, palpebral ptosis and a decrease in motor activity. Autopsy results showed no apparent change in lungs, heart, liver, spleen and kidneys except for changes in the weight of some organs (Saleem et al. 1999).

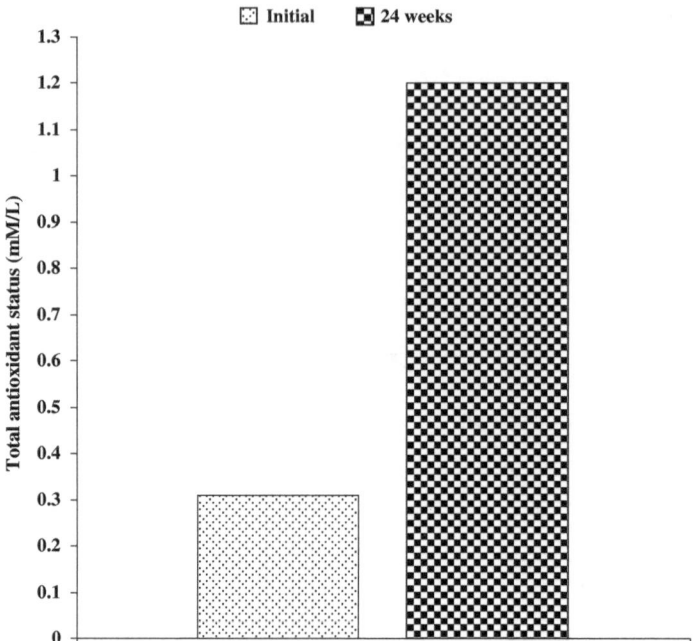

Fig. 4.5 Remarkable improvement in total antioxidant status after administration of *B. ceiba* root powder in a patient of involuntary weight loss (Verma et al. 2011)

Saleem et al. (2003) have further studied acute toxicology of most active hypotensive fraction of methanolic extract of defatted stem bark of *B. ceiba* in NMR-1 mice by oral administration in two sets of animals. One received 100 mg/kg/day and other 1000 mg/kg/day for 7 days. At a dose of 100 mg/kg/day it caused expiry of two male and one female mice and male mice showed 17% loss in body weight while 7% weight loss was observed in female mice. At 1000 mg/kg/day, all the animals were expired and it proved to be a lethal dose (LD_{100}). Histopathology of heart, kidney and liver showed damaging effect of fraction of methanolic extract of stem bark. Interestingly, oral administration of BCBMM at a dose of 200 mg/kg/day did not show any mortality in Sprague-Dawley rats.

Dar et al. (2005) have studied effect of mangiferin on strychnine-induced lethality in NMR-1 mice. Control group received 10% DMSO (10 ml/kg orally) followed after 1 h by oral administration of strychnine (1 mg/kg). Group two was treated similarly except that mangiferin (10 mg/kg; orally) was given instead of DMSO. At the sublethal dose of strychnine, no mortality was observed in mangiferin treated mice showing that mangiferin is not an inhibitor of microsomal drug metabolizing enzyme.

Acute toxicity of aqueous extract of *B. ceiba* flowers through oral and I.V. route was also assessed in rats and mice. It was found that the magnitude and integrity of the toxic symptoms was highly dose-dependent. The LD_{50} for oral route in rats was 6768.730 mg/kg and for I.V. route it was 889.496 and 467.84 mg/kg in rats and mice, respectively (Rehman et al. 2006). Recently, Said et al. (2011) have shown that oral administration of 70% methanolic extract of flowers up to 5 g/kg body weight induced no obvious toxic effects in mice and all animals remained alive after 24 h.

Singh et al. (2001) studied extracts of pollen and seed fibers of *B. ceiba* in type-1 allergic patients. Pollen extract has markedly showed positive skin sensitivity in 5.6% nasobronchial allergy patients out of total 105 patients. Old and new Silk-Cotton fiber extract elicited positive skin sensitivity in 12.4 and 6.6% of total patients, respectively. Pollen extract showed 15 allergenic proteins on immunoblot out of which proteins of 50, 42, 35, 30, 20 and 14 kDa molecular weight were detected as major IgE binding proteins.

4.17 Conclusion

Silk-cotton tree is a plant full of health-promoting phytopharmaceuticals, of extreme importance with medicinal properties and a strong ethnobotanical background. Ethnomedicine-based research has revealed it to be a plant having anti-hyperglycemic, antihyperlipidaemic, antimicrobial, antihypertensive, antioxidant, fibrinolysis-enhancing and anabolic potential. Further researches are needed to explore these pharmacological activities for newer drug development.

References

Aratanechemuge Y, Hibasami H, Sanpin K, Katsuzaki H, Imai K, Komiya T (2004) Induction of apoptosis by lupeol isolated from mokumen (*Gossampinus malabarica* L. Merr) in human promyelotic leukemia HL-60 cells. Oncol Rep 11(2):289–292

Bhargava C, Thakur M, Yadav SK (2011) Effect of *Bombax ceiba* L. on spermatogenesis, sexual behaviour and erectile function in male rats. Andrologia. doi:10.1111/j.1439-0272.2011.01210.x

Chen J, Zhuang D, Cai W, Xi L, Li E, Wu Y, Sugiama K (2009) Inhibitory effects of four plants flavonoids extracts on fatty acid synthase. J Envir Sci Suppl 21:S131–S134

Chiu HF, Lin CC, Yen MH, Wu PS, Yang CY (1992) Pharmacological and pathological studies on hepatic protective crude drugs from Taiwan (V): The effects of *Bombax malabarica* and *Scutellaria rivularis*. Am J Chinese Med 20(3–4):257–264

Chowdhuri S, Chatterjee PC (1973) Survey of the haemagglutinating properties of plant seeds and fungi. Indian J Med Res 61:1478–1484

Dar A, Faizi S, Naqvi S, Roome T, Zikr-ur-Rehman S, Ali M, Firdous S, Moin ST (2005) Analgesic and antioxidant activity of mangiferin and its derivatives: the structure activity relationship. Biol Pharm Bull 28(4):596–600

Dhawan BN, Saxena PN (1958) Evaluation of some indigenous drugs for stimulant effect on the rat uterus: a preliminary report. Indian J Med Res 46(6):808–811

Dhar ML, Dhar MM, Dhawan BN, Mehrotra BN, Ray C (1968) Screening of Indian plants for Biological property: Part I. Indian J Exp Biol 6:232–247

Faizi S, Ali M (1999) Shamimin: a new flavonol C-glycoside from leaves of *Bombax ceiba*. Planta Med 65(4):383–385

Gupta A, Gupta R (1997) A survey of plants for presence of cholinesterase activity. Phytochemistry 46:827–831

Gupta RA, Singh BN, Singh RN (1981) Preliminary study of certain vednasthapana (analgesic) drugs. J Sci Res Plant Med 2(4):110–112

Hossain E, Chandra G, Nandy AP, Mandal SC, Gupta JK (2011a) Anthelmintic effect of a methanol extract of *Bombax malabaricum* leaves on *Paramphistomum explanatum*. Parasitol Res. doi:10.1007/s00436-011-2594-y

Hossain E, Rawani A, Chandra G, Mandal SC, Gupta JK (2011b) Larvicidal activity of *Dregea volubilis* and *Bombax malabaricum* leaf extracts against the filarial vector *Culex quinquefasciatus*. Asian Pac J Trop Med 4(6):436–441

Hossain E, Mandal SC, Gupta JK (2011c) Phytochemical screening and *in vivo* antipyretic activity of the methanol leaf-extract of *Bombax malabaricum* DC (Bombacaceae). Trop J Pharm Res 10(1):55–60

Islam MK, Chowdhury JA, Eti IZ (2011) Biological activity study on a malvaceae plant: *Bombax ceiba*. J Sci Res 3(2):445–450

Jagtap AG, Niphadkar PV, Phadke AS (2011) Protective effect of aqueous extract of *Bombax malabaricum* DC on experimental models of inflammatory bowel disease in rats and mice. Indian J Exp Biol 49(5):343–351

Jain V (2009) Isolation of active principles and effect of crude drugs obtained from *Ipomoea digitata* Linn. and *Bombax ceiba* Linn. for their antioxidant property vis-à-vis endothelial dysfunction in human beings. Ph.D. Thesis, Department of Botany, Mohanlal Sukhadia University, Udaipur

Jain V, Verma SK, Katewa SS, Anandjiwala S, Singh B (2011a) Free radical scavenging property of *Bombax ceiba* root. Res J Med Plant 5(4):462–470

Jain V, Katewa SS, Verma SK (2011b) *In vitro* antimicrobial activity of roots of *Bombax ceiba*— An ethnomedicinal plant. Proceedings of international conference on folk and herbal medicine, Scientific Publishers, Jodhpur

Kong FY, Ng DK, Chan CH, Yu W-L, Chan D, Kwok K-L, Chow P-Y (2006) Parental use of the term "Hot Qi" to describe symptoms in their children in Hong Kong: a cross sectional survey "Hot Qi" in children. J Ethnobiol Ethnomed 5:2

Lin CC, Chen SY, Lin JM, Chiu HF (1992) The pharmacological and pathological studies on Taiwan folk medicine (VIII): the anti-inflammatory and liver protective effects of "Mu-mien". Am J Chinese Med 20(2):135–146

Mishra DN, Dixit V, Mishra AK (1991) Mycotoxic evaluation of some higher plants against ringworm causing fungi. Indian Drugs 28:300–303

Misra MB, Mishra SS (1966) Studies in indigenous enterotonic drugs (a preliminary note). Indian J Physiol Pharmacol 10:59

Misra MB, Mishra SS, Mishra RK (1968) Pharmacology of *Bombax malabaricum* DC. Indian J Pharmacy 30:165

Nakahara K, Roy MK, Alzoreky NS, Thalang VN, Trakoontivakorn G (2002) Inventory of indigenous plants and minor crops in Thailand based on bioactivities. In: 9th JIRCAS international symposium—"Value-addition to Agricultural Products", 135–139

Nam NH, Kim HM, Bae KH, Ahn BZ (2003) Inhibitory effects of Vietnamese medicinal plants on tube-like formation of human umbilical venous cells. Phytother Res 17(2):107–111

Patel SS, Verma NK, Rathore B, Nayak G, Singhai AK, Singh P (2011) Cardioprotective effect of *Bombax ceiba* flowers against acute adriamycin-induced myocardial infarction in rats. Rev Bras Farmacogn. doi:10.1590/S0102-695X2011005000090

Qi YP, Zhu H, Guo SM, Jue HQ, Lin S (2008) Study on antitumor activity of extract from roots of *Gossampinus malabaria*. Zhong Yao Cai 31(2):266–268

Rani P, Khullar N (2004) Antimicrobial evaluation of some medicinal plants for their anti-enteric potential against multi-drug resistant *Salmonella typhi*. Phytother Res 18(8):670–673

Ravi V, Patel SS, Verma NK, Dutta D, Saleem TSM (2010) Hepatoprotective activity of *Bombax ceiba* Linn against Isoniazid and Rifampicin-induced toxicity in experimental rats. Int J Appl Res Nat Prod 3(3):19–26

Rehman Z, Rehman A, Ahmad S (2006) Acute toxicity studies of *Bombax ceiba* flowers in mice and rats. Pak J Sci Ind Res 49(6):410–413

Said A, Aboutable EA, Nofal SM, Tokuda H, Raslan M (2011) Phytoconstituents and bioactivity evaluation of *Bombax ceiba* L. flowers. J Trad Med 28:55–62

Saleem R, Ahmad M, Hussain SA, Qazi AM, Ahmad SI, Qazi MH, Ali M, Faizi S, Akhtar S, Hussain SN (1999) Hypotensive, hypoglycaemic and toxicological studies on the Flavonol C-glucoside Shamimin from *Bombax ceiba*. Planta Med 65:331–334

Saleem R, Ahmad SI, Ahmad M, Faizi Z, Rehman S, Ali M, Faizi S (2003) Hypotensive activity and toxicology of constituents from *Bombax ceiba* stem bark. Biol Pharm Bull 26(1):41–46

Shahat AA, Hassan RA, Nazif NM, Van Miert S, Pieters L, Hammuda FM, Vlietinck AJ (2003) Isolation of mangiferin from *Bombax malabaricum* and structure revision of shamimin. Planta Med 69(11):1068–1070

Sharma M, Mishra SS (1969) A pharmacological study of some abortifacient plants (a preliminary report). Indian J Physiol Pharmacol 13:139–141

Shyur LF, Tsung JH, Chen JH, Chiu CY, Lo CP (2005) Antioxidant properties of extracts from medicinal plants popularly used in Taiwan. Int J Appl Sci Eng 3(3):195–202

Singh BP, Verma J, Shridhara S, Rai D, Gaur SN, Arora N (2001) Allergens of *Salmalia malabarica* (Eng. Silk Cotton) tree pollen and seed fibres. Indian J Allergy Appl Immunol 15(1):45–48

Somanadham B, Varughese G, Palpu P, Sreedharan R, Gudiksen L, Smitt UW, Nyman U (1999) An ethnopharmacological survey for potential angiotensin converting enzyme inhibitors from Indian medicinal plants. J Ethnopharmacol 65:103–112

Surveswaran S, Cai YZ, Corke H, Sun M (2007) Systematic evaluation of natural phenolic antioxidants from 133 Indian medicinal plants. Food Chem 102:938–953

Verma SK, Jain V, Katewa SS (2006) Fibrinolysis enhancement by *Bombax ceiba*—a new property of an old plant. South Asian J Prev Cardiol 10(4):212–219

Verma SK, Jain V, Katewa SS (2008) Potential antihyperglycemic activity of *Bombax ceiba* in type 2 diabetes. Int J Pharmacol Biol Sci 2(1):79–86

Verma SK, Jain V, Katewa SS (2011) Anabolic effect of *Bombax ceiba* root in idiopathic involuntary weight loss—a case study. J Herb Med Toxicol 5(1):1–5

Vieira TO, Said A, Aboutabl E, Azzam M, Creczynski-Pasa TB (2009) Antioxidant activity of methanolic extract of *Bombax ceiba*. Redox Rep 14(1):41–46

Wang YC, Huang TL (2005) Screening of anti-*Helicobacter pylori* herbs deriving from Taiwanese folk medicinal plants. FEMS Immunol Med Microbial 1:43(2):295–300

Wiyakrutta S, Sriubolmas N, Panphut W, Thongon N, Danwisetkanjana K, Ruangrungsi N, Meevootisom V (2004) Endophytic fungi with anti-microbial, anti-cancer and anti-malarial activities isolated from Thai medicinal plants. World J Microbiol Biotechnol 20:265–272

Xu J, Huang Z, Li C et al. (1993) Anti-inflammatory activity of alcohol extract from flower of *Gossampinus malabarica*. J Fujian Medical University 2. doi: cnki:ISSN:1672-4194.0.1993-02-005

You YJ, Nam NH, Kim HM, Bae KH, Ahn BZ (2003) Antiangiogenic activity of lupeol from *Bombax ceiba*. Phytother Res 17(4):341–344

Yu YG, He QT, Yuan K, Xiao XL, Li XF, Liu DM, Wu H (2011) *In vitro* antioxidant activity of *Bombax malabaricum* flower extracts. Pharm Biol 49(6):569–576

Chapter 5
Commercial Importance

Abstract Commercially, floss, wood and seed oil of Red Silk-Cotton tree are important. Timber is mainly used in matchstick production, vermin-proof floss for stuffing upholstery and life-saving appliances and seed oil is utilized as edible oil and for soap-making.

Keywords Indian kapok · Timber · Kapok-oil · Silk-cotton · Indian cotton-wood

5.1 Introduction

Silk-cotton tree, the name clearly depicts the soft silk (Fig. 2.2m–o) obtained from its fruits which is used widely for various purposes. Logs of *B. ceiba* are very important for matchstick industries as well as for fishermen. This requires a constant supply of logs and thus puts this plant species under great pressure (Baral et al. 2005). Leaves and young twigs are lopped as fodder. Gum (Fig. 2.2f–h) of the tree, mixed with ashes and castor oil, is used as cement for caulking iron pans for boiling sugar (Chadha 1972). Audu-Peter et al. (2009) has shown that the gum isolated from the calyx of *B. ceiba* possesses interesting physico-chemical properties and can be used for pharmaceutical purposes.

5.2 Floss

Fruits of *B. ceiba* are a good source of silk-cotton which comprises the Indian Kapok of commerce. The weight of 100 dry fruits is about 2 kg of which floss is 600 g and seeds are 500 g. These brownish-yellow fibers originate from the inner side of the capsule and are light, buoyant, elastic, resilient and water-repellent in nature. It loses only 10% of its buoyancy after remaining in water for 30 days and after fully drying, it regains the buoyancy and resilience. Length of fibers ranges from 17.78 to 27.94 mm and diameter varies between 0.01524 and 0.03392 mm. It contains 61–64% cellulose, 2.7% ash and less than 25% pentosan. A full bearing *B. ceiba* tree produces 4.5–6.0 kg of clean floss free from seeds, whereas true kapok tree produces about 2 kg. For obtaining better quality floss, semi-ripe fruits are collected directly from trees and sun-dried for 4–6 days. Floss is then scooped out and further sun-dried for 10–12 days. It is cleaned and ginned by means of power-driven machines and recovery of ginned floss ranges between 35 and 40% (by weight) of the unginned floss (Chadha 1972; Ghose 1942).

Inner stem bark of *B. ceiba* also yields a fiber suitable for cordage and making twine (Brock 1988). Kapok, the fiber isolated from its fruits is used to fill mattresses, life preservers and sleeping bags, is an important non-wood forest product (Online document 1 2006). Floss is used for stuffing life-belts, life-saving appliances, sleeping bags, mattresses, cushions, pillows, upholstery, wadded cloth quilts etc. Floss can be dry sterilized at 110°C without any loss of its properties, so it is an excellent material for making padded surgical dressings. Waxy coating of floss makes it vermin-proof and suitable for stuffing cushions, pillows and upholstery. Floss is also used as an insulating material for refrigerators, soundproof covers and walls and it is superior to cotton wool as packing material for fragile goods. Nowadays it is spun into yarns and used for manufacture of plushes and laces (Chadha 1972; Pandey 2005).

5.3 Wood

Wood of *B. ceiba* is usually white or pale-pink and turns pale-yellowish brown on exposure, often streaked with sapstain or decay. It is straight-grained, even and coarse textured, and very light and soft. Timber is air or kiln seasoned and sterilized for 2 h at 55°C and 100% relative humidity. Timber of this plant is most widely used in match industry and good for boxes as well as splints. Indian cottonwood requires a 40–50-year rotation to grow large enough to produce quality veneer, which exacerbates the supply problem in match industry due to its declining status in various states of India (Tandon JC 2011).

It is also used for planking ceilings, canoes, catamarans, shingles, toys, pencils, pen-holders, veneers, scabbards, coffins, well curbs, brush-handles, picture frames, ladles (*Chatu*), as cushions for mine-props, inside partition of opium-chests, paper pulp (Jain et al. 2009; Choudhary et al. 2008; Pandey 2005) and wooden sculptures, didgeridoos and artifact production (Griffiths et al. 2003; Whitehead et al. 2006) etc. The wood is widely used for preparation of canoes and catamarans; however, it is rapidly destroyed by marine borers. Tarakanadha et al. (2006) have shown that copper biocides such as ammoniacal copper zinc arsenate and chromated copper arsenate provide protection to its wood from marine borers.

Plant is known as *Bulamana* among aboriginals of Mannigrada community, northern territory, Australia and they use its timber widely for arts and crafts industry (Whitehead et al. 2006). *Kathodi* tribe of Rajasthan make use of its wood for preparing musical instruments such as a membranophonic *Dholak* and *Tambura* (Joshi 1995); while *Bhil* tribe use its wood to make spoons for their kitchen (Singh and Pandey 1998).

Preferably, wood is used for loose or slack cooperages, which are used for dry goods such as cement. Timber of the plant is strong, elastic, durable and free from defects and that is why it is the most recommended species for ship and boat building (Pandey 2005). Plywood made from this wood is weaker than commercial plywoods; but still it is used for making tea-boxes, fruit-crates and packing cases. Fine shavings of the wood are known as wood-wool which is used as a good packing material and for making cement-bounded wood-wool boards (Chadha 1972).

5.4 Oil

Oil isolated from its seeds is comparable to true Kapok and can be used for edible purposes as a substitute for cotton-seed oil, for soap-making and as an illuminant. Oil cake is richer in protein than that of true Kapok and cottonseed cakes and is an excellent cattle feed as it contains very little or no gossypol, the toxic principle of cotton-seed (Chadha 1972).

References

Audu-Peter JD, Vandi JK, Wuniah B (2009) Evaluation of some physicochemical properties of *Bombax* gum. J Pharm Bioresources 6(2):38–42

Baral N, Gautam R, Tamang B (2005) Population status and breeding ecology of White-rumped Vulture *Gyps bengalensis* in Rampur Valley, Nepal. Forktail 21:87–91

Brock J (1988) Top end native plants. J. Brock, Darwin

Chadha YR (1972) The Wealth of India, raw material, vol IX. Publications and Information Directorate, New Delhi Vol. IX

Choudhary BL, Katewa SS, Galav PK (2008) Plants in material culture of tribal and rural communities of Rajsamand district of Rajasthan. Indian J Trad Knowl 7(1):11–22

Ghose TP (1942) Indian kapok. Indian Forest Leafl 32:1–10

Griffiths AD, Phillips A, Godjuwa C (2003) Harvest of *Bombax ceiba* for the aboriginal arts industry, Central Arnhem Land, Australia. Biol Cons 113(2):295–305

Jain V, Verma SK, Katewa SS (2009) Myths, traditions and fate of multipurpose *Bombax ceiba*—an appraisal. Indian J Trad Knowl 8(4):638–644

Joshi P (1995) Ethnobotany of primitive tribes in Rajasthan. Printwell, Jaipur

Online document 1. International trade in non-wood forest products in the Asia pacific region. http://www.fao.org//docrep/x5336e/x5336e0f.htm. Accessed 14 May 2006

Pandey BP (2005) Economic botany. S. Chand and Company Limited, New Delhi

Singh V, Pandey RP (1998) Ethnobotany of Rajasthan, India. Scientific Publishers, Jodhpur

Tandon JC 4.0 Case study two, The safety match industry in India. FAO corporate document repository. http://www.fao.org/docrep/X5860E/x5860e05.htm. Accessed 12 Nov 2011

Tarakanadha B, Rao KS, Narayanappa P, Morrell JJ (2006) Marine performance of *Bombax ceiba* treated with inorganic preservatives. J Trop Forest Sci 18(1):55–58

Whitehead PJ, Gorman J, Griffiths AD, Wightman G, Massarella H, Altman J (2006) Feasibility of small scale commercial native plant harvests by indigenous communities. Rural Industries Research and Development Corporation, Kingston Act

Chapter 6
Ecological Importance and Need of Conservation

Abstract Silk-Cotton tree is in fact a boon to environment by providing shelter and food to animals, reducing pollution, cooling homes, attracting positive microvita and thus plays an important role in balancing ecosystem. However, few traditions and commercial overexploitation is leading to an untimely demise of this multipurpose tree species in various parts of the world. Results of some conservation strategies applied in Udaipur, India have also been discussed.

Keywords *Holi* · Agroforestry · Vermi-compost · Pollution · SMRIM · Ex situ conservation

6.1 *Bombax ceiba*: A Multipurpose Tree Species

Bombax ceiba is not only useful medicinally and economically but is also an equally important plant species for ecosystem. It is considered as a reforestation pioneer plant species and survives easily in low rainfall and well-drained conditions. It also coppices well in the early stages of growth. Plants can be raised by direct seed sowing, entire transplanting and stump-planting methods. Seeds are embedded in soft, silk cotton and usually there are 25,300–38,500 seeds in 1 kg of floss. Light weight of floss helps in easy dispersal of seeds through blowing wind. Propagation is also done with branch-cuttings, but the survival rate is poor in this method (Chadha 1972).

The plant is well known for its beautiful crimson red flowers, which are used to make natural color for *Holi*—the festival of colors in India. Water extract of this flowers has been found to be useful as a natural acid–base indicator safer than synthetic indicators. Thus, this eco-friendly natural indicator can be used to minimize chemical pollution as well as to provide safety to users (Soltan and Sirry 2002).

V. Jain and S. K. Verma, *Pharmacology of Bombax ceiba Linn.*,
SpringerBriefs in Pharmacology and Toxicology,
DOI: 10.1007/978-3-642-27904-1_6, © The Author(s) 2012

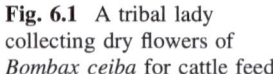

Fig. 6.1 A tribal lady
collecting dry flowers of
Bombax ceiba for cattle feed

Silk-Cotton tree is also known for its fire resistance and soil binding property (Chadha 1972; D'Silva 2001). Recently, it has been shown that good quality of vermicompost can be prepared from its dry leaves and flowers. Dry leaves took 8 months to convert in vermicompost while flowers took only 3.5 months. Vermicompost developed from leaves and flowers contained total (%) N 1.4, 1.2; P 0.6, 0.55; K 2.1, 1.77; Na 2.3, 1.4; Ca 4.25, 0.45 and S 1.4, 0.9, respectively (Sannigrahi 2009). It has been demonstrated that this plant possesses moderate SO_2 absorption efficacy and therefore, can be recommended for roadside plantations in polluted urban areas (Farooq et al. 1988).

In silvi-pastoral system of Agroforestry, protein-rich multipurpose trees are planted around farmlands and rangelands for cut and carry fodder production to meet the feed requirements of livestock during the fodder deficit period in winter (Fig. 6.1). *Bombax ceiba* is the most preferred fodder species due to its high protein content and considered as an important multipurpose tree for providing food, fodder, fuel, timber and soil protection (Chundawat and Gautam 1993).

The Red Indian Kapok tree is an important medicinal plant species which has been shown to be useful for reclamation of wastelands and mine spoils (Dadhwal and Singh 1993). This multipurpose tree species can be used for social forestry schemes and for planting in barren areas to improve the soil and to gain economic benefits, species propagation and soil conservation. It is a large ornamental tree, acts as a living fence and can be widely planted for urban greening (Online document 1 2011).

Trees provide shade and insulate the home from heat loss or gain and thus, can be used around the home for passive energy conservation. This passive energy saving landscaping concept for climate control is known as enviroscaping. The energy-conserving impact of a particular tree species depends on whether it keeps its leaves during the winter and the shape of a tree and density of its foliage. Medium shade density, large spreading nature and high drought tolerance makes Silk-Cotton Tree one of the most suitable tree species for passive energy conservation around the homes as studied in South Florida (Broschat et al. 2007).

Fig. 6.2 a–g Dependence of various animal species on *Bombax ceiba*

Bright crimson red flowers appear in very large numbers on the tree from January to March and make the whole tree a mass of brilliant color. To enhance its beauty, many mynas and rosy pastors chat lively along with bass of crows and twittering of sparrows and other small birds who visit the tree in search of nectar (Santapau 1996) and therefore, this tree merits consideration as a keystone resource (Raju et al. 2005). Being predominantly an ornithophilous plant species, it is also designated as *'Little bird's cafeteria'*. It serves as a nesting site for Rock bees, Red-Weaver Ants and birds such as Coppersmith Barbets, Indian Gray Hornbills, Vultures and Greater Adjutant Storks. Its nectar-rich flowers are used as a source of food by many animals such as Hanuman Langurs, Peacocks, Squirrels, Blue bull, Indian Crested Porcupine, House Crow, Sunbirds, Red-Vented Bulbul, House sparrow, Common Myna, Black Drongo, Parakeets, Indian Roller Bird, *Dysdercus* bugs and *Glenea cantor*. (Fig. 6.2a–g). Not only this, but also the nectar present in its flowers also possess microscopic mycoflora. Mushtaq et al. (2006) have isolated total 16 fungal species belonging to ten genera from the nectar. These were *Bullera megalospora, Candida rhagii, C. sake, C. valdiviana, Debaryomyces hansenii, Fibulobasidium inconspicuum, Filobasidiella neoformans,*

Pichia castillae, P. dryadoides, P. lynferdii, P. mississippiensis, P. ohmeri, Pseudozyma antarctica, Stephanoascus ciferrii, Williopsis californica and *Zygoascus helinicus*. Thus, *B. ceiba* is an excellent example of '*Umbrella tree species*' and should be recommended for plantation in tropical forests, protected areas as well as home gardens to further support lives of many animal species that are dependent on it (Bhatt and Kumar 2001; Jain et al. 2011; Kohno and Buithi 2005; Liu et al. 2007; Lu et al. 2011).

Unfortunately, decline in number of *B. ceiba* trees had shown a great impact on many animals surviving on it in various parts of the world. A study on population status and breeding ecology of *Gyps bengalensis* (White-rumped Vulture) in Rampur valley, Nepal, has shown that 86% nests of *G. bengalensis* were built on *B. ceiba* trees. Commercial logging of *B. ceiba* trees in Nepal has limited its nest-site availability leading to drastic reduction in population of White-rumped Vultures which are nature's best scavengers (Baral et al. 2005). Similarly, decline in population of an endangered bird *Leptoptilos dubius* (Greater Adjutant Stork) has also been observed in Assam, India which is due to continuous reduction in the availability of its nesting sites in general and overexploitation of *B. ceiba* in particular as it is the most preferred nesting site of Greater Adjutant Stork (Singha et al. 2002).

In Maradavally forest area of Western Ghats, India, *B. ceiba* has been the main shelter tree for honeybee colonies. But overexploitation of this plant species has put it into an endangered species and also resulted in reduction of honeybee populations which are the main pollinating agents. The loss of pollinators has further induced reduction in the seed set of *B. ceiba*. Therefore, depletion of such keystone resource species within a forest area can cause a deleterious impact on the whole ecosystem (Ramkrishnappa and Krell 2006).

6.2 The Need of Conservation

The plant is being exploited largely for medicinal and commercial purposes but traditional burning of this tree in *Holika-dahan*, which is an important festival of North India, is jeopardizing its survival in the tribal-dominated Udaipur district of Rajasthan, India (Verma et al. 2006; Sharma 2006; Jain et al. 2007). Commercial overexploitation has also raised questions on its survival in Nepal and Australia (Baral et al. 2005; Whitehead et al. 2006).

Various ethnoconservation practices, in the form of traditions, customs, myths and folk-tales, have made survival of *B. ceiba* on the earth for so many years. The in situ conservation practices of indigenous communities can be considered as the traditional ecological heritage, which conserves the population of varied plant species in its habitat; the best method of conservation. Now there is a need to revive these beneficial traditions of conservation to protect many endemic and endangered plant species as well as to protect the habitat for future generations. The present case study deals with status of *B. ceiba* and various conservation efforts made to protect this multipurpose tree species in Udaipur, India.

6.2.1 Case Study

Udaipur district is situated between 24°34'48" north latitude and 73°40'48" east longitude. It is particularly a tribal-dominated division, where *Meena, Bhil, Garasia, Damor* and *Kathodi* are the main inhabiting tribes. Among the various customs and traditions related to *B. ceiba*, the one which is the most dangerous and widely prevalent even in urban masses of Udaipur city is the use of this tree as a pole for *Holika-dahan* (Jain et al. 2007, 2009). Due to this particular prevalent tradition of cutting and burning *Semal* tree in huge quantities in a single day, every year, the plant is now available sparingly in few patches at Kewara ki Naal, Desuri ki Naal, Chirwa valley and Hawala village in the vicinity of Udaipur city. Our continuous survey since last 5 years (2006–2011) has shown many illegal tree felling events in nearby forest areas of the Udaipur city just before *Holi* festival. Many trees were badly mutilated and some cut logs were lying in such position, so that they can be easily transported in the night for selling in urban areas (Fig. 6.3).

Enquiry from the local tribes and people who were openly selling the tree at various places in and around the city and from elderly people of the community in different areas of the city, involved in *Holi* burning events, gave us an astonishing fact about the number of trees which were sacrificed for this festival (Fig. 6.4). In the year 2007, around 2,000–2,500 trees or twigs of *Semal* tree were cut for *Holika-dahan* and sold in the city at the rates of Rs. 200–300 per *Holi*-pole. The cost of *Holi*-poles depending on the length and breadth of the plant was increased up to Rs. 500 per pole in the year 2009–2010. However, in the year 2011, the cost was significantly decreased to Rs. 150 and even there was a significant reduction observed in the number of *Semal* trees brought for selling.

Besides this, another devastating tradition of burning the longest *Holi* of the city is jeopardizing survival of this plant. From the year 2007 onwards till 2011, a 54–55 ft long *Semal* tree, brought from nearby forest area, at the cost of around Rs. 3,000–4,000 is being burnt at Subhash Chawk, Mullatalai, Udaipur without any concern about the sacrifice of such a large tree (Fig. 6.5).

Tradition of cutting and burning *Semal* trees from last many years has led to reduction of *B. ceiba* to such an extent that tribals are now cutting other plant species and selling them in the name of *Semal* tree. It shows how an irrational traditional practice can become so devastating and it is obvious that loss of even one part of flora or fauna of food chain can disturb the ecosystem drastically.

6.2.2 Conservation Efforts

Looking to the grave situation of *B. ceiba* in Udaipur district, a strategy was prepared as '*Semal Conservation Mission*' under Society for Microvita Research and Integrated Medicine (SMRIM), Udaipur to combat with this social evil leading to untimely death of such an important plant species. It included both in situ and

Fig. 6.3 a–f Badly mutilated
trees of *B. ceiba* to be used
for burning in *Holika-dahan*

ex situ conservation efforts. For this, the first step was to create awareness among
masses and to involve them in both in situ and ex situ conservation programs.

6.2.3 Awareness Campaign

An intensive awareness program was initiated since 2006 and is still continuing.
It included publication about the grave situation of *B. ceiba* in Udaipur in various
national and international journals, magazines and newspapers along with pam-
phlet distribution among masses with an urge to conserve the plant for future
(Verma et al. 2006; Jain et al. 2007, 2009; Verma and Jain 2008a, b; Jain and
Verma 2008; Jain 2009, 2010a, b, c, 2011a).

Fig. 6.4 a–f Author interacting with the sellers of debarked stems of *B. ceiba* in Udaipur, India on the occasion of *Holi* festival

Another important effort in the direction of awareness was direct interactions with various rural and urban communities (both young and old generations) in the form of group meetings, table talks, workshops and lecture deliveries (Jain 2011b). After 5 years of intensive efforts, support was provided by a renowned local newspaper '*Rajasthan Patrika*' in 2011, who agreed to collaborate in our *Semal Conservation Mission* program. It helped us to materialize our newly proposed concept of burning an iron pole as a *Holi-danda* instead of *B. ceiba* or any other plant species (Verma 2011).

6.2.4 In Situ Conservation

Results of *Semal conservation mission* became clearly evident in 2011. With the involvement of forest officials, who were initially not supporting the mission in name of conserving the tradition, also became active and did not allow any kind of illegal felling of *Semal* trees in the forests. This really helped the mission as there

Fig. 6.5 Burning of longest
Holi of Udaipur. **a** Attempts
to raise the longest *Holi*.
b Decorated longest *Holi*
ready for burning. **c** Silk
cotton tree in flames and
fumes of tradition

were very few *Holi-dandas* seen in the market in 2011 as compared to 2007 and
that too were sold at much reduced cost.

To reduce the burden on forests and for real protection of *Semal* trees in the
wild, an innovative idea was proposed by the authors Verma and Jain (2008b).
It was suggested that an iron pole wrapped with dried grass and hay material
instead of the wooden pole of *B. ceiba* can also be used for burning in *Holi* where
the grass will be burnt to ashes while iron pole will remain unharmed, just as the
Prahlad survived in the fire in the mythological story. So, there will be no need to
sacrifice *B. ceiba* or any other tree on this occasion.

This idea was ultimately materialized in March 2011 and we were quite
successful infusing this idea into minds of urban masses. Members of SMRIM,
Udaipur prepared Iron-*Holi* poles and provided it to local people as per demand
(Fig. 6.6). Members of SMRIM burnt this eco-friendly *Holi* covered in fire-proof
cotton at Arvind Nagar, Sunderwas and well organized this event by showing the
people importance of such beneficial medicinal plant (Figs. 6.7, 6.8). Many edu-
cated communities came forward and adopted this Iron-pole concept keeping
behind all dogmatic thinking which were perpetuating till then. Forest officials also
burnt an iron *Holi* at all their forest divisions showing the importance of *Semal* tree
to tribal communities. This was a real reward that came from local people in
direction of *Semal* conservation.

The awareness attempts also led us to another success in reducing the numbers
of *Holi* burning events. For this, group meetings were conducted with influential
persons of various communities and the positive result was that all of them agreed

Fig. 6.6 Iron poles as an alternate to *Semal* for burning *Holi*

Fig. 6.7 SMRIM members preparing *Holi* by covering iron pole with hay material

Fig. 6.8 Burning of iron pole as *Holi*

to burn a Group-*Holi* and thus avoided unnecessary cutting of *Semal* trees in large quantities. They also agreed to plant five *Semal* saplings every year for every tree cut during *Holika-dahan*. This was a rewarding step toward *Semal* conservation strategies after such rigorous 5 years continuous efforts which proved fruitful at the end.

6.2.5 Ex Situ Conservation

It is the most desirable objective with medicinal plants. It is more to offer than systematic storage of germplasm in botanical gardens, field gene banks and seed collections. For this, seeds of the plant are being collected every year in the month of April since 2007 from wild by SMRIM and are being sown in favorable environmental conditions at SMRIM's nursery, Tekri-Madri Road, Udaipur (Fig. 6.9a–c).

Till July 2011, around 600 saplings have been recovered and about 500 have already been transplanted in and around Udaipur city. The places where *Semal*

Fig. 6.9 a–b Saplings of *Bombax ceiba*. **c** Propagation of *Semal* in nursery managed by SMRIM. **d–g** Plantation of *Semal* at various places in Udaipur, India

plants have been transplanted are as follows: Neemuch Mata Hill, Doodh Talai Lake area, Chitrakoot Nagar, RNT Medical College, Gulab Bagh Zoo, Sajjangarh Wildlife Sanctuary, University College of Science, College of Law, Mohanlal Sukhadia University, Sukhadia Samadhi Garden, Manohar Vatika, Sunderwas, Rameshwar Park, Sector no. 4, Hiranmagari, Moti Magri garden and museum compound near Fatehsagar lake, J. C. Bose Hostel, Rajasthan College of Agriculture, Physiotherapy, Homeopathy and Lokmanya Tilak B.Ed. College, Janardan Rai Nagar Vidyapeeth University, Dabok, Khaddeshwar Mahadev temple near Airport, Amrakh Mahadev temple, near Chirwa valley, District Collector's residence at Court circle, Rajiv Gandhi Garden, Adinda Parshwanath Temple, College of Technology and Engineering, Maharana Pratap University of Agriculture Campus, Udaipur and in various Primary and Middle schools at Bhinder municipality area (Fig. 6.9d–g). It is worth noting here that there was not a single plant of *Semal* present at all these places except at the Botanical garden, University College of Science, Mohanlal Sukhadia University.

Till date *Semal* plant has been conserved and propagated only by nature and to some extent by the indigenous communities in form of various ethnoconservation practices. Due to such intensive awareness program running for past 5 years, many intellectuals came forward and allowed to plant *Semal* in their own houses and are really taking care of them properly. Propagation by developing the plants from seeds and transplanting them on such a large scale in Udaipur is a historical step which will surely revitalize the importance of this medicinal plant species as well as prove a milestone in the direction of *Semal* conservation. Efforts are being continued for its propagation and preservation both in situ and ex situ.

6.2.6 Conclusion

These large-scale ex situ conservation efforts for a tree of ethnomedicinal value and spiritual importance is a rewarding step in the history of Udaipur which will surely prove to be a clarion call to nature lovers, environmentalists and the general public to come forward and help in conservation of this plant species so that it may be saved from an untimely demise. Re-establishment of ethnoconservation practices and traditions, which are not causing harm to the environment, is the dire need of time (see Chap. 2). Similar approaches can be made in other countries also to save this plant species while simultaneously utilizing it on a sustainable basis.

References

Baral N, Gautam R, Tamang B (2005) Population status and breeding ecology of White-rumped Vulture *Gyps bengalensis* in Rampur Valley. Nepal Forktail 21:87–91

Bhatt D, Kumar A (2001) Foraging ecology of red-vented Bulbul *Pycnonotus cafer* in Haridwar. India Forktail 17:109–110

Broschat TK, Meerow AW, Black RJ (2007) EES-42. Enviroscaping to conserve energy: trees for South Florida. http://edis.ifas.ufl.edu/EH142.htm. Accessed 23 Apr 2008

Chadha YR (1972) The wealth of India, raw material. Vol. IX. Publications and Information Directorate, New Delhi

Chundawat BS, Gautam SK (1993) Textbook of agroforestry. Oxford and IBH Publishing Co. Pvt. Ltd, New Delhi

Dadhwal KS, Singh B (1993) Trees for the reclamation of limestone mine spoil. J Ind Soc Soil Sci 41(4):738–744

D'Silva E (2001) Community forest management in Adilabad district, Andhra Pradesh, India, Ecological effects. Asia Forest Network, Working paper series, Santa Barbara, California, vol. 4, 2nd ed. www.asiaforestnetwork.org. Accessed 28 Nov 2006

Farooq M, Saxena RP, Beg MU (1988) Sulfur dioxide resistance of Indian trees-I. Experimental evaluation of visible symptoms and SO_2 sorption. Water Air Soil Pollut 40(3–4):307–316

Jain V (2009) Holi ki agni mei swaha Semal ke Aansu (In Hindi). BOMRIM 1(1):3

Jain V (2010a) Arogyadata Semal (Poem in Hindi). Bappa Rawal 1(5):21

Jain V (2010b) Bahudeshiya Semal Vriksha (In Hindi). Anuvrat 56(3):16–17

Jain V (2010c) Dhanatmak microvita sampann bahudeshiya Semal vriksha (In Hindi). Bappa Rawal 1(6):25–27

Jain V (2011a) Dev vriksha-Semal (Poem in Hindi). Bhakta Samaja 10(4):33

Jain V (2011b) Journey of Semal research and conservation. BOMRIM 3(1):2–3

Jain V, Verma SK (2008) Conservation of silk cotton tree. Prout 19(8):22–23

Jain V, Verma SK, Katewa SS (2007) A dogmatic tradition posing threat to *Bombax ceiba*—the Indian Red Kapok tree. Medicinal Plant Conserv 13:12–15

Jain V, Verma SK, Katewa SS (2009) Myths, traditions and fate of multipurpose *Bombax ceiba*— an appraisal. Indian J Tradit Knowl 8(4):638–644

Jain V, Verma SK, Sharma SK, Katewa SS (2011) *Bombax ceiba* Linn.—As an umbrella tree species in forests of southern Rajasthan, India. Res J Environ Sci 5(8):722–729

Kohno K, Buithi N (2005) Comparison of the life history strategies of three *Dysdercus* true bugs (Heteroptera: Pyrrhocoridae), with special reference to their seasonal host plant use. Entomol Sci 8(4):313–322

Liu F, Roubik DW, He D, Li J (2007) Old comb for nesting site recognition by *Apis dorsata*? Field experiments in China. Insect Soc. doi:10.1007/s00040-007-0963-4

Lu W, Wang Q, Tian MY, Xu J, Qin AZ (2011) Phenology and laboratory rearing procedures of an Asian longicorn beetle, *Glenea cantor* (Coleoptera: Cerambycidae: Lamiinae). J Econ Entomol 104(2):509–516

Mushtaq M, Jamal A, Nahar S (2006) Biodiversity of yeast mycoflora in nectar of *Bombax ceiba* and *Canna indica* flower. Pak J Bot 38(4):1279–1288

Online document 1. Malabulak (*Bombax ceiba*)—indigenous to the Philippines and other countries (13.6.2011). http://www.indi-journal.info/archives/4192. Accessed 28 Sept 2011

Raju AJS, Rao SP, Rangaiah K (2005) Pollination by bats and birds in the obligate outcrosser *Bombax ceiba* L. (Bombacaceae), a tropical dry season flowering tree species in the Eastern Ghats forests of India. Ornithological Sci 4:81–87

Ramakrishnappa A, Krell R (2006) Case study no. 8. Impact of cultivation and gathering of medicinal plants on biodiversity: case studies from India. FAO corporate document repository. http://www.fao.org/DOCREP/005/Y4586E/y4586e09.htm. Accessed 10 May 2006

Sannigrahi AK (2009) Biodegradation of leaf litter of tree species in presence of cow dung and earthworms. Indian J Biotechnol 8:335–338

Santapau H (1996) Chapter 7: silk-cotton tree. In: Santapau H (ed) Common trees. National Book Trust, New Delhi

Sharma SK (2006) Wildlife management. Himanshu Publications, Udaipur

Singha H, Rahmani AR, Coulter MC, Javed S (2002) Nesting ecology of the greater adjutant stork in Assam, India. Waterbirds 25(2):214–220

Soltan ME, Sirry SM (2002) Usefulness of some plant flowers as natural acid-base indicators. J Chin Chem Soc 49:63–68

Verma SK (2011) Iron Holi-alternate to semal—a historical step (In Hindi). BOMRIM 3(1):1

Verma SK, Jain V (2008a) Silk cotton tree in flames and fumes of tradition. Prout 19(1):18–19

Verma SK, Jain V (2008b) Holi ki agni mei dhadhakata Dev Vriksha-Semal (In Hindi). Pratyush 6(2):14–15

Verma SK, Jain V, Katewa SS (2006) Urgent need for conservation of silk cotton tree (*Bombax ceiba*)—a plant of ethno medicinal importance. Bull Biol Sci 4(1):81–84

Whitehead PJ, Gorman J, Griffiths AD, Wightman G, Massarella H, Altman J (2006) Feasibility of small scale commercial native plant harvests by indigenous communities. Rural Industries Research and Development Corporation, Kingston Act

Chapter 7
Future Research

Abstract *Bombax ceiba* Linn is historically, psycho-spiritually, commercially, ecologically and medicinally important plant species. Phytochemically it is enriched with flavonoids, phenolics and sesquiterpenoids. Fragmented research work has been done in various parts of the world to validate its ethno-medicinal properties and the results are outstanding. However, it still needs detailed exploration of its beneficial biological potential against hyperlipidemia, hyperglycemia and atherosclerosis.

Keywords Antiviral · Hypotensive · Hypoglycemic · Anti-inflammatory · Atherosclerosis

Bombax ceiba, the Indian Red Kapok tree is an important multipurpose tree species and has been proved to be a nature's boon for human welfare. The flowers are beautiful bright red and full of energy, the stem bark is resistant to fire; protecting the plant and the forest; the floss from the seeds is used for stuffing, timber is used for match and art industries, the gum is used for female health and the root for treating diabetes and male sexual disorders. It have a very high reputation in ethnomedicine and traditional medicinal systems which has been scientifically proved in various experimental studies. However, extensive research should be carried out in the following directions:

- Every part of the plant has been subjected for phytochemical analysis resulting in isolation of some novel compounds such as Bombamaloside, Bombamalones, Bombasin, Bombalin and Shamimicin. However, further research should go with a pace to isolate the active components with the proved biological activities such as antiviral, antimutagenic, hypoglycemic, hypolipidemic, androgenic and anabolic.
- It has shown some potential against free radicals, lipids, COX pathway and coagulation parameters. Future research should be aimed toward detailed

exploration of its action on prevention/regression of experimentally induced atherosclerosis.

- Large-scale, placebo-controlled, double-blind, multi-centered, cross-over studies with a large number of patients for long duration should be carried out to establish its hypolipidemic, hypoglycemic, antioxidant, anabolic and fibrinolysis-enhancing activities.
- Incorporation of hypoglycemic and hypolipidemic components from root, hypotensives from leaves/flowers and antioxidants from flowers/root may pave way for production of a 'Herbal Polypill' which may prove useful for the presently prevailing epidemics of coronary heart disease and diabetes.
- Extracts prepared from its various parts should be standardized with the help of chromatography and spectroscopic techniques for quality assurance and control.
- Looking to its immense therapeutic potential and important role in ecosystem, its declining population is a serious threat to humanity. Therefore, there is an urgent need to develop ex situ conservation techniques such as tissue culture and micropropagation to propagate this plant species in short time.

In a nutshell, *Bombax ceiba* is a plant having a long life span and providing ecological benefits. There is a need for further research to assess its bio-medicinal values. Moreover, the plant is used for so many purposes that countries such as India, Nepal and Australia have started facing the increased rate of harvesting, which may threaten future sustainability of the plant. Therefore, there is urgent need to develop large-scale in situ and ex situ conservation techniques and their implementation for preserving this important plant species.

Index

A
Abortion, 14, 20
ABTS, 52
Acne, 8, 13, 14, 22
Agroforestry, 73–74, 86
Anabolic, 8, 51, 60–61, 65, 67, 89–90
Analgesic, 8, 12, 42–45, 47, 51, 56, 65
Anemia, 14
Angiotensin, 62, 67
Anthelmintic, 55, 66
Anti-angiogenic, 42, 44, 57
Antibacterial, 9, 34, 43–45, 48–49, 54
Antidiabetic, 14, 21, 43, 60
Anti-fertility, 14
Anti-inflammatory, 8, 12, 22, 25, 34, 43–44,
    47, 55–56, 66–67, 89
Antimicrobial, 43, 51, 53–54, 65
Antimutagenic, 9, 42, 57, 89
Antinociceptive, 44
Antioxidant, 9, 24, 34, 43–44, 51–53,
    61, 64, 90
Antipyretic, 57
Antiviral, 43, 54, 89
Aphrodisiac, 1, 8–9, 13, 60
Apigenin, 29–30, 35–36, 43
Artifact, 70
Aspergillus niger, 55
Asthma, 13
Astringent, 8–9
Atherosclerosis, 89–90

B
Bacillus subtilis, 34, 54
Bee, 18–19, 75–76

Beta-sitosterol, 26–28, 31, 34,
    37, 42
Bird, 75–76
Black Drongo, 75
Blood pressure, 42, 59
Blue bull, 75
Boat, 71
Bombacaceae, 1–2, 57
Bombamalone, 25–26, 34, 37, 89
Bombamaloside, 25–26, 34, 37, 89
Bombaxquinone, 26, 34, 37
Bulamana, 71
Bullera megalospora, 75
Burn, 1, 8, 13–14, 16, 80

C
Cancer, 14, 34, 42–44, 57
Candida albicans, 34, 54–55
Candida rhagii, 75
Candida sake, 75
Candida valdiviana, 75
Ceiba pentandra, 44, 57–58
Chicken pox, 14
Cholinesterase, 44, 62
Chrysanthemum morifolium, 9
Colitis, 8, 14, 56
Common Myna, 75
Conjunctivitis, 13, 14
Conservation, 17–20, 73–74, 77–80,
    83, 85, 90
Constipation, 61
Coppersmith Barbet, 75
Cordia gharaf, 15
Cosmetin, 30, 35, 43

V. Jain and S. K. Verma, *Pharmacology of Bombax ceiba Linn.*,
SpringerBriefs in Pharmacology and Toxicology,
DOI: 10.1007/978-3-642-27904-1, © The Author(s) 2012